Civil Engineering: A Very Short Introduction

VERY SHORT INTRODUCTIONS are for anyone wanting a stimulating and accessible way in to a new subject. They are written by experts, and have been published in more than 25 languages worldwide.

The series began in 1995 and now represents a wide variety of topics in history, philosophy, religion, science, and the humanities. The VSI library now contains more than 300 volumes—a Very Short Introduction to everything from ancient Egypt and Indian philosophy to conceptual art and cosmology—and will continue to grow in a variety of disciplines.

Very Short Introductions available now:

ADVERTISING Winston Fletcher
AFRICAN HISTORY John Parker and Richard Rathbone
AGNOSTICISM Robin Le Poidevin
AMERICAN IMMIGRATION David A. Gerber
AMERICAN POLITICAL PARTIES AND ELECTIONS L. Sandy Maisel
THE AMERICAN PRESIDENCY Charles O. Jones
ANAESTHESIA Aidan O'Donnell
ANARCHISM Colin Ward
ANCIENT EGYPT Ian Shaw
ANCIENT GREECE Paul Cartledge
ANCIENT PHILOSOPHY Julia Annas
ANCIENT WARFARE Harry Sidebottom
ANGELS David Albert Jones
ANGLICANISM Mark Chapman
THE ANGLO-SAXON AGE John Blair
THE ANIMAL KINGDOM Peter Holland
ANIMAL RIGHTS David DeGrazia
THE ANTARCTIC Klaus Dodds
ANTISEMITISM Steven Beller
ANXIETY Daniel Freeman and Jason Freeman
THE APOCRYPHAL GOSPELS Paul Foster
ARCHAEOLOGY Paul Bahn
ARCHITECTURE Andrew Ballantyne
ARISTOCRACY William Doyle
ARISTOTLE Jonathan Barnes
ART HISTORY Dana Arnold
ART THEORY Cynthia Freeland
ATHEISM Julian Baggini
AUGUSTINE Henry Chadwick
AUSTRALIA Kenneth Morgan
AUTISM Uta Frith
THE AZTECS David Carrasco
BARTHES Jonathan Culler
BEAUTY Roger Scruton
BESTSELLERS John Sutherland
THE BIBLE John Riches
BIBLICAL ARCHAEOLOGY Eric H. Cline
BIOGRAPHY Hermione Lee
THE BLUES Elijah Wald
THE BOOK OF MORMON Terryl Givens
BORDERS Alexander C. Diener
THE BRAIN Michael O'Shea
BRITISH POLITICS Anthony Wright
BUDDHA Michael Carrithers
BUDDHISM Damien Keown
BUDDHIST ETHICS Damien Keown
CANCER Nicholas James
CAPITALISM James Fulcher
CATHOLICISM Gerald O'Collins
THE CELL Terence Allen and Graham Cowling
THE CELTS Barry Cunliffe
CHAOS Leonard Smith
CHILDREN'S LITERATURE Kimberley Reynolds
CHINESE LITERATURE Sabina Knight
CHOICE THEORY Michael Allingham
CHRISTIAN ART Beth Williamson
CHRISTIAN ETHICS D. Stephen Long
CHRISTIANITY Linda Woodhead
CITIZENSHIP Richard Bellamy
CIVIL ENGINEERING David Muir Wood
CLASSICAL MYTHOLOGY Helen Morales

CLASSICS Mary Beard and John Henderson
CLAUSEWITZ Michael Howard
THE COLD WAR Robert McMahon
COLONIAL LATIN AMERICAN LITERATURE Rolena Adorno
COMMUNISM Leslie Holmes
THE COMPUTER Darrel Ince
THE CONQUISTADORS Matthew Restall and Felipe Fernández-Armesto
CONSCIENCE Paul Strohm
CONSCIOUSNESS Susan Blackmore
CONTEMPORARY ART Julian Stallabrass
CONTINENTAL PHILOSOPHY Simon Critchley
COSMOLOGY Peter Coles
CRITICAL THEORY Stephen Eric Bronner
THE CRUSADES Christopher Tyerman
CRYPTOGRAPHY Fred Piper and Sean Murphy
THE CULTURAL REVOLUTION Richard Curt Kraus
DADA AND SURREALISM David Hopkins
DARWIN Jonathan Howard
THE DEAD SEA SCROLLS Timothy Lim
DEMOCRACY Bernard Crick
DERRIDA Simon Glendinning
DESCARTES Tom Sorell
DESERTS Nick Middleton
DESIGN John Heskett
DEVELOPMENTAL BIOLOGY Lewis Wolpert
THE DEVIL Darren Oldridge
DICTIONARIES Lynda Mugglestone
DINOSAURS David Norman
DIPLOMACY Joseph M. Siracusa
DOCUMENTARY FILM Patricia Aufderheide
DREAMING J. Allan Hobson
DRUGS Leslie Iversen
DRUIDS Barry Cunliffe
EARLY MUSIC Thomas Forrest Kelly
THE EARTH Martin Redfern
ECONOMICS Partha Dasgupta
EGYPTIAN MYTH Geraldine Pinch
EIGHTEENTH-CENTURY BRITAIN Paul Langford
THE ELEMENTS Philip Ball
EMOTION Dylan Evans
EMPIRE Stephen Howe
ENGELS Terrell Carver
ENGINEERING David Blockley
ENGLISH LITERATURE Jonathan Bate
ENVIRONMENTAL ECONOMICS Stephen Smith
EPIDEMIOLOGY Rodolfo Saracci
ETHICS Simon Blackburn
THE EUROPEAN UNION John Pinder and Simon Usherwood
EVOLUTION Brian and Deborah Charlesworth
EXISTENTIALISM Thomas Flynn
FASCISM Kevin Passmore
FASHION Rebecca Arnold
FEMINISM Margaret Walters
FILM Michael Wood
FILM MUSIC Kathryn Kalinak
THE FIRST WORLD WAR Michael Howard
FOLK MUSIC Mark Slobin
FORENSIC PSYCHOLOGY David Canter
FORENSIC SCIENCE Jim Fraser
FOSSILS Keith Thomson
FOUCAULT Gary Gutting
FREE SPEECH Nigel Warburton
FREE WILL Thomas Pink
FRENCH LITERATURE John D. Lyons
THE FRENCH REVOLUTION William Doyle
FREUD Anthony Storr
FUNDAMENTALISM Malise Ruthven
GALAXIES John Gribbin
GALILEO Stillman Drake
GAME THEORY Ken Binmore
GANDHI Bhikhu Parekh
GENIUS Andrew Robinson
GEOGRAPHY John Matthews and David Herbert
GEOPOLITICS Klaus Dodds
GERMAN LITERATURE Nicholas Boyle
GERMAN PHILOSOPHY Andrew Bowie
GLOBAL CATASTROPHES Bill McGuire
GLOBAL ECONOMIC HISTORY Robert C. Allen
GLOBAL WARMING Mark Maslin
GLOBALIZATION Manfred Steger
THE GOTHIC Nick Groom
THE GREAT DEPRESSION AND THE NEW DEAL Eric Rauchway

NEWTON Robert Iliffe
NIETZSCHE Michael Tanner
NINETEENTH-CENTURY BRITAIN Christopher Harvie and H. C. G. Matthew
THE NORMAN CONQUEST George Garnett
NORTH AMERICAN INDIANS Theda Perdue and Michael D. Green
NORTHERN IRELAND Marc Mulholland
NOTHING Frank Close
NUCLEAR POWER Maxwell Irvine
NUCLEAR WEAPONS Joseph M. Siracusa
NUMBERS Peter M. Higgins
OBJECTIVITY Stephen Gaukroger
THE OLD TESTAMENT Michael D. Coogan
ORGANIZATIONS Mary Jo Hatch
PAGANISM Owen Davies
PARTICLE PHYSICS Frank Close
PAUL E. P. Sanders
PENTECOSTALISM William K. Kay
THE PERIODIC TABLE Eric R. Scerri
PHILOSOPHY Edward Craig
PHILOSOPHY OF LAW Raymond Wacks
PHILOSOPHY OF SCIENCE Samir Okasha
PHOTOGRAPHY Steve Edwards
PLAGUE Paul Slack
PLANETS David A. Rothery
PLANTS Timothy Walker
PLATO Julia Annas
POLITICAL PHILOSOPHY David Miller
POLITICS Kenneth Minogue
POSTCOLONIALISM Robert Young
POSTMODERNISM Christopher Butler
POSTSTRUCTURALISM Catherine Belsey
PREHISTORY Chris Gosden
PRESOCRATIC PHILOSOPHY Catherine Osborne
PRIVACY Raymond Wacks
PROBABILITY John Haigh
PROGRESSIVISM Walter Nugent
PROTESTANTISM Mark A. Noll
PSYCHIATRY Tom Burns
PSYCHOLOGY Gillian Butler and Freda McManus
PURITANISM Francis J. Bremer
THE QUAKERS Pink Dandelion
QUANTUM THEORY John Polkinghorne
RACISM Ali Rattansi
RADIOACTIVITY Claudio Tuniz
THE REAGAN REVOLUTION Gil Troy
REALITY Jan Westerhoff
THE REFORMATION Peter Marshall
RELATIVITY Russell Stannard
RELIGION IN AMERICA Timothy Beal
THE RENAISSANCE Jerry Brotton
RENAISSANCE ART Geraldine A. Johnson
RISK Baruch Fischhoff and John Kadvany
RIVERS Nick Middleton
ROBOTICS Alan Winfield
ROMAN BRITAIN Peter Salway
THE ROMAN EMPIRE Christopher Kelly
THE ROMAN REPUBLIC David M. Gwynn
ROMANTICISM Michael Ferber
ROUSSEAU Robert Wokler
RUSSELL A. C. Grayling
RUSSIAN HISTORY Geoffrey Hosking
RUSSIAN LITERATURE Catriona Kelly
THE RUSSIAN REVOLUTION S. A. Smith
SCHIZOPHRENIA Chris Frith and Eve Johnstone
SCHOPENHAUER Christopher Janaway
SCIENCE AND RELIGION Thomas Dixon
SCIENCE FICTION David Seed
THE SCIENTIFIC REVOLUTION Lawrence M. Principe
SCOTLAND Rab Houston
SEXUALITY Véronique Mottier
SHAKESPEARE Germaine Greer
SIKHISM Eleanor Nesbitt
SLEEP Steven W. Lockley and Russell G. Foster
SOCIAL AND CULTURAL ANTHROPOLOGY John Monaghan and Peter Just
SOCIALISM Michael Newman
SOCIOLOGY Steve Bruce
SOCRATES C. C. W. Taylor
THE SOVIET UNION Stephen Lovell
THE SPANISH CIVIL WAR Helen Graham

SPANISH LITERATURE Jo Labanyi
SPINOZA Roger Scruton
STARS Andrew King
STATISTICS David J. Hand
STEM CELLS Jonathan Slack
STUART BRITAIN John Morrill
SUPERCONDUCTIVITY Stephen Blundell
TERRORISM Charles Townshend
THEOLOGY David F. Ford
THOMAS AQUINAS Fergus Kerr
TOCQUEVILLE Harvey C. Mansfield
TRAGEDY Adrian Poole
TRUST Katherine Hawley
THE TUDORS John Guy
TWENTIETH-CENTURY BRITAIN Kenneth O. Morgan
THE UNITED NATIONS Jussi M. Hanhimäki
THE U.S. CONGRESS Donald A. Ritchie
THE U.S. SUPREME COURT Linda Greenhouse
UTOPIANISM Lyman Tower Sargent
THE VIKINGS Julian Richards
VIRUSES Dorothy H. Crawford
WITCHCRAFT Malcolm Gaskill
WITTGENSTEIN A. C. Grayling
WORLD MUSIC Philip Bohlman
THE WORLD TRADE ORGANIZATION Amrita Narlikar
WRITING AND SCRIPT Andrew Robinson

Available soon:

GOVERNANCE Mark Bevir
THE ORCHESTRA D. Kern Holoman
WORK Stephen Fineman
AMERICAN HISTORY Paul S. Boyer
SPIRITUALITY Philip Sheldrake

For more information visit our website
www.oup.com/vsi/

David Muir Wood

CIVIL ENGINEERING

A Very Short Introduction

OXFORD
UNIVERSITY PRESS

Great Clarendon Street, Oxford, OX2 6DP,
United Kingdom

Oxford University Press is a department of the University of Oxford.
It furthers the University's objective of excellence in research, scholarship, and education by publishing worldwide. Oxford is a registered trade mark of Oxford University Press in the UK and in certain other countries

First Edition published in 2012

Impression: 10

British Library Cataloguing in Publication Data
Data available

Library of Congress Cataloging in Publication Data
Data available

ISBN 978–0–19–957863–4

Printed in Great Britain by
Ashford Colour Press Ltd, Gosport, Hampshire

Acknowledgements

Early versions of the text of this book have been read by Margaret Abel, Hugh Balchin, Carol Collins, Julia Elton, Ted and Virginia Khan, Adrian Mathias, Helen Muir Wood, James Sutherland, and Sue Vardy. I am immensely grateful to all of them, but especially to Julia and James, for the thoughtful and detailed comments that they provided. Whether they will feel that I have adequately reacted to their suggestions I cannot tell – the responsibility for the final text rests with me.

I am also very grateful to Andrew Muir Wood for providing excellent drawings for Figures 10, 18, 24 and 27.

David Muir Wood
Monikie
April 2012

Contents

List of illustrations

Introduction

In 1828, the Royal Charter of the Institution of Civil Engineers, drafted by Thomas Tredgold, defined the purpose of the Institution:

> *The general advancement of mechanical science, and more particularly for promoting the acquisition of that species of knowledge which constitutes the profession of a civil engineer; being the art of directing the great sources of power in nature for the use and convenience of man, as the means of production and of traffic in states, both for external and internal trade, as applied in the construction of roads, bridges, aqueducts, canals, river navigation, and docks, for internal intercourse and exchange; and in the construction of ports, harbours, moles, breakwaters, and light-houses, and in the art of navigation by artificial power, for the purposes of commerce; and in the construction and adaptation of machinery, and in the drainage of cities and towns.*

Setting off on a camping trip in the hills the associated needs are evident. You need shelter from the elements and from potential predators. You need a supply of fresh water, food, and energy for cooking, heating, and light. There is some advantage in thinking ahead about 'waste disposal' and possibly the need for medical supplies. Wandering the wilderness has its attraction but more frequently there will be some form of pathway in existence for you to follow. The pioneering people reaching the Americas or other unfamiliar lands had the same requirements. Interdependence encouraged the formation of settlements which became villages and then towns and cities. The benefits of scale and collaboration in provision of this infrastructure were obvious—common means of protection against predators (human or animal), common sources of clean water (no need for each household to have its own well), common strategies for disposal of waste (one man's waste disposal can become another man's pollution), common sources of energy, common networks of transport. The advantages of sharing skills and resources would have rapidly become apparent.

All these elements of infrastructure come under Tredgold's definition of 'civil engineering'.

The emergence of civil engineering

The term *civil engineering* has been used, and its meaning has evolved for rather more than 200 years. Chambers Dictionary defines an engineer as 'one who designs or makes, or puts to practical use, engines or machinery of any type, including electrical; one who designs or constructs public works, such as roads, railways, sewers, bridges, harbours, canals, etc; one who constructs or manages military fortifications, etc., or engines'. A civil engineer is 'one who plans and builds railways, docks, etc. as opposed to a military engineer, or to a mechanical engineer, who makes machines etc.'

'Engineer' comes from the Latin *ingenium* meaning skill, linked with the design and construction of clever devices (*engines*)—such as catapults—for inflicting damage in military campaigns. The term *civil engineering*, introduced to distinguish those working with non-military engines, seems to be almost an oxymoron. But the term now relates to that branch of engineering which is concerned with the creation of the infrastructure of society: the *civil* implying this link with the citizen and with civilization. The techniques used for these *civil* projects have general application: civil engineers may well be working with the military on provision of buildings, roads, bridges, airfields—specifically *un-civil* engineering under the original meaning of the term.

Of course, even before the term *civil engineering* was coined, civil engineers were in existence even if they were then described using other terms. The person (or persons) unknown who organized the construction of the Great Pyramids, some 5000 years ago, must be called an engineer. He (probably a man) certainly had acquired, by observation and experience, the skills necessary to translate a concept into reality using effective technology. The Greeks (Anthemius of Thrales and Isidorus of Miletus) put in charge of the design of the church of Hagia Sophia (in Constantinople – Byzantium or Istanbul) by the Roman Emperor Justinian in around AD 600 were, like Christopher Wren 1000 years later, expected both to solve the day-to-day problems of detailed construction and also to plan the overall shape and form: they were civil engineers. The Greek word μηχανή (mechani) means 'a machine, or a clever trick'. Odysseus had the attribute πολυμήχανος (polimechanos): 'one who knew many tricks and could cope with difficult situations'. This would be a good description of an engineer, but today a mechanic would be expected to have a narrower range of skills than an engineer. Anthemius and Isidoros would have been described as αρχιτέκτονες (architektones): *master builders*. The potential for confusion of engineering and architecture is apparent. The inventiveness of cathedral builders such as Brunelleschi in devising special cranes for safely raising construction supplies would have qualified them for the description of *engineer* too. Words like *mechanic*, *engineer*, *architect* now resonate in a way which does not begin to think about their origins two or three millennia ago.

Much mediaeval engineering was based on geometric rules based on prior experience of successful structures. Scientific understanding developed slowly as people such as Galileo sought explanations for phenomena and began to apply mathematical concepts to the analysis of structural elements. The Industrial Revolution which began in Britain in the early 18th century produced both a large number of ingenious steam-powered machines, which replaced man-power in the new factories, and also a new building material—cast iron—which had a strength greater than the timber that it replaced in many bridges and large buildings. With care in the manufacturing process, there was a predictability and narrow variation of the mechanical properties which made its use more straightforward.

The cathedral builders (or engineers) organized themselves into a closed community, keeping the secrets of their experience within the community. The notion of engineering as a profession is a development from this but it occurred in different ways in different countries. The French, with an emphasis on Cartesian analysis and mathematical representation, made rapid progress in what we now see as the various branches of engineering mechanics, with state encouragement. The British, with an emphasis on observation and practice, and less desire for central control, developed a more autonomous profession. The national routes to the *formation* of engineers differ widely in the balances of practice and theory or of training and education.

In late 17th century France, the military engineer was given formal status by the creation of a *Corps des Ingénieurs du Génie Militaire* in 1690 by Maréchal Vauban. Vauban was famous for the design of fortifications, which usually consisted of masonry structures retaining compacted earth (within the remit of civil engineering today). In 1716, this was followed by the creation of the *Corps des Ingénieurs des Ponts et Chaussées* who were charged with the construction and maintenance of networks of communication (roads and bridges). The training of these latter engineers was formalized with the creation in 1747 of the *École des Ponts et Chaussées*, the first of the French *Grandes Écoles*: the concept of civil engineering as distinct from military engineering was already clear in France. The first director of the *École des Ponts et Chaussées* was Jean-Rodolphe Perronet who began his career with responsibility for the design of Paris sewers and roads – civil engineering infrastructure.

In Britain, the Royal Society – founded in 1661 by Charles II – brought together individuals who shared a curiosity in observation of phenomena. Some took part in the search for scientific explanations of what they had seen. Christopher Wren and Robert Hooke were both members and worked together on the reconstruction of London after the fire of 1666. The boundary between what we might now see as science and engineering was rather fuzzy. We remember Wren as an architect and designer of buildings, Hooke more for his scientific insight (though Newton, who succeeded him as President of the Royal Society, did all he could to obliterate his memory). Both Wren and Hooke could equally be described as engineers.

John Smeaton, born in 1724, was elected a fellow of the Royal Society in 1753, in recognition of his work on scientific instruments (which we would now categorize as mechanical engineering – he went on to study water and windmills). His ability to combine scientific interests with a skill in design resulted in his being recommended for the design of the Eddystone Lighthouse off the south-west coast of England. This success then developed into a major career in civil engineering, designing and supervising the construction of bridges, roads, and canals across England and Scotland. He was the first person to describe himself as a *civil* engineer, deliberately to distinguish himself from the military engineers, and was the natural person to create, in 1771, a Society of Engineers which met over dinner to share engineering problems and solutions. The membership of this rather informal group identified three classes of members: real engineers, actually employed as such; gentlemen of rank and fortune who have applied their minds to civil engineering; and various artists whose professions are useful to civil engineering: quite enlightened in its breadth.

Smeaton's Society of Engineers was the precursor of the Institution of Civil Engineers, which was founded in a coffee house on the Strand, in London, in January 1818 by three young engineers – Henry Palmer, James Jones, and Joshua Field. It did not have much influence as a professional engineering organization until Thomas Telford was elected President in 1820. He was well known for his civil engineering prowess and had built a network of political and society contacts. He also knew most of those who were actively practising civil engineering in Britain. It was his drive which resulted in the award of the Royal Charter for the Institution in 1828, thus establishing the Institution as the visible professional body for engineers of all disciplines. (The Smeatonian Society of Civil Engineers remains in existence as an exclusive dining club and claims to be the oldest engineering society in the world.)

The 19th century was a time of revolutions in Europe, a birth of new countries, and an upsurge of nationalism and self-confidence as industrialization and international commerce brought increased prosperity to nations and expansion to cities. It was inevitable that schisms should similarly develop in the newly established Institution of Civil Engineers. The Institution of Mechanical Engineers was created in 1847 (under the presidency of George Stephenson the railway engineer whose son and engineering partner Robert became President of the Institution of Civil Engineers in 1855). The (Royal) Institution of Naval Architects was founded in 1860 by (among others) John Scott Russell who had worked with Brunel on the *Great Eastern* (as partner or opponent, depending on which account you believe) and who was forced to resign his vice-presidency of the Institution of Civil Engineers because of his alleged involvement in some arms deals during the American Civil War. The Royal Aeronautical Society was founded in 1866. The Society of Telegraph Engineers (subsequently to become the Institution of Electrical Engineers) was founded in 1871. The Institution of Municipal Engineers was founded in 1874 with the encouragement of the Institution of Civil Engineers, with which they amalgamated in 1984. The Concrete Institute (subsequently the Institution of Structural Engineers) was founded in 1908. And so on... The sphere of influence of civil engineering was whittled away and although attempts have been made to merge some of these institutions in order to improve their national and international influence and to recognize the interdisciplinary nature of many of the most challenging engineering projects, the extent of consolidation in the UK has been somewhat limited. Some other countries have either avoided the schisms or have successfully reintegrated the profession.

Today, the term 'civil engineering' distinguishes the engineering of the provision of infrastructure from these many other branches of engineering that have come into existence. It thus has a somewhat narrower scope now than it had in the 18th and early 19th centuries. There is a tendency to define it by exclusion: civil engineering is not mechanical engineering, not electrical engineering, not aeronautical engineering, not chemical engineering... But that is to lose sight of what is included.

In the 19th century, William Rankine was typical of those engineers/scientists who had a broad interest in phenomena and the theory to describe them. He is known today both for his developments in thermodynamics and the underpinning theory of heat engines, and for key theoretical analysis of the pressure exerted on earth retaining structures – enabling the design of walls supporting slopes or terraces, for example. Was he a civil engineer? Yes, of course. Was he a mechanical engineer? Well, yes, but the concept did not really exist until the middle of the century. Similarly Charles Coulomb is known by schoolchildren for his work on electric charge but is also well known to students of civil engineering for his theoretical work on the pressure of earth on retaining structures. Was he a civil engineer? His work on earth pressures was performed while he was in the French army in Martinique and designing defensive earthworks as part of the fortifications. So by strict definition he was not a *civil* engineer although the work for which he was responsible would today be seen as falling within the remit of a civil engineer. His work on electrostatics also makes him an electrical engineer.

As a footnote to this discussion of the nature and definition of civil engineering, it is interesting to note that in some countries – for example, Norway and Sweden – the term civil engineer has retained something of its earlier broad meaning as an indicator of a level of educational attainment in an engineering field. One might be a civil engineer with qualification in electrical engineering or in information technology, or in chemical engineering – or in building engineering or in road and water construction. There is always room for diversity and confusion.

Civil engineering today is seen as encompassing much of the infrastructure of modern society provided it does not move – roads, buildings, dams, tunnels, drains, airports (but not aeroplanes or air traffic control), railways (but not railway engines or signalling), power stations (but not turbines). The fuzzy definition of civil engineering as the engineering of infrastructure, the lubrication of society, should make us recognize that there are no precise boundaries and that any practising engineer is likely to have to communicate across whatever boundaries appear to have been created. This is a challenge for education to ensure that the *languages* of boundary disciplines are not so disparate that communication is prevented. The boundary with science is also fuzzy. Engineering is concerned with the solution of problems *now*, and cannot necessarily wait for the underlying science to catch up. But as scientific understanding of the bases for engineering decisions improves, so also does the confidence level attached to those decisions.

All engineering is concerned with finding solutions to problems for which there is rarely a single answer. Presented with an appropriate 'solution-neutral problem definition' the engineer needs to find ways of applying existing or emergent technologies to the solution of the problem. There are a number of distinctive features which add excitement to the practice of civil engineering: civil engineering projects tend to be large, they tend to be visible, and they tend to be unique. At every stage through selection of structural form, choice of materials, to final appearance, decisions have to be made. There is a subjectivity, of course, but also an immense feeling of pride in having been closely involved with the decision making process which led to the construction of *that* bridge, *that* tunnel, *that* building. People are thus an important element of civil engineering projects. Though there may be possibilities for mass production of structural elements such as beams and columns or even sections of bridge decks, the structures into which they are incorporated will usually be quite different. The corollary is that, like the coke-can ring-pull, the project has got to work first time. The size and visibility mean that failures cannot be hidden.

The uniqueness of most civil engineering projects means that, while theory and analysis certainly have their place, direct experience or experience transmitted by older engineers remains important: '*good judgement comes from experience, experience comes from bad judgement*'. The dramatic increase in the power and availability of computers over the past 50–60 years has opened up possibilities for (approximate) numerical solutions to engineering problems which could not have been contemplated previously. Those computer analyses require as input the output of scientific studies of material response: science and engineering proceed hand in hand. Decisions formerly made on the basis of experience ('*would a structure built with these proportions be likely to stand up*?') can now be supported by analysis. However, it is too easy to be taken in by the first colourful outpourings of a computer program without confirming, by simple back-of-the-envelope calculation, drawing on experience or simplified modelling, that the answers are at least of the correct order of magnitude.

Architecture and civil engineering

Inevitably many of the example projects that will be described here will be structures – for example, buildings and bridges. These are among the more visible creations of civil engineers, but we tend to associate architects rather than engineers with such structures – though originally an architect was simply a 'master builder' (or engineer). The distinctive separation of the professions has occurred gradually over the past few centuries.

Sir Christopher Wren is remembered as an architect of the second half of the 17th century, yet in his activities in designing and supervising the construction of St Paul's Cathedral after the Great Fire of London he was as much an engineer (in our 21st century interpretation of the term) as an architect. The distinction did not then exist in Britain. Projects such as the creation of channels to supply water to growing conurbations required understanding of the mechanics of pumps and mills and steam engines. The emphasis of the projects fell firmly onto engineers (to handle all the engines) and surveyors, and the opportunity for involvement of architects (as we understand the term) was limited.

In early 18th century France, the *Corps du Génie* was concerned with the (civil) engineering of defence fortifications for withstanding the siege of towns. The *Corps des Ponts et Chaussées* was responsible for the (civil) engineering of roads, canals, water supply. A third institution, the *Bâtiments du Roi*, was linked with aesthetics of engineered buildings as required by the king. In America, the Corps of Engineers, though primarily concerned with matters of defence and fortification in time of war, manoeuvred itself into a powerful position, reclaiming land from mudflats on the Potomac river and developing the new capital city of Washington DC.

The introduction of iron as a construction material after the Industrial Revolution produced a breed of fabricator/contractor, developing skills onto a larger scale from the work of the pre-existing blacksmith. In dealings with architects they had a clear notion of the capabilities and limitations of their material. Gustave Eiffel was primarily an enterprising contractor but had sufficient numeracy and design skills to be thought of now as an engineer and designer. The rapidly increasing extent of scientific knowledge could be applied to the analysis of structural elements and systems (and the ground). Enlightened programmes of education tried to maintain a breadth which kept engineering and architecture together. However, increasing investment in roads and railways and building (and other elements of infrastructure) led to increasing demands for specialist engineering designers.

Andrew Saint suggests that a 'broadbrush way of describing the change in relations between architects and engineers over the past two centuries might be to say that they used to work on different projects but have similar skills, whereas they now work on the same projects but have different skills,' and also, 'Architecture lacks the linearity of engineering, that quality of consistent serviceability and of the creative skill subordinating itself to the practical end at issue... It is architecture's freedom from the manipulations of utility that makes it a true art'. Many of the attempts to separate engineering and architecture seem to be based on rhetoric rather than on the actual process of creating structures. 'Once we enter into what happens when a structure is actually assembled in any age, we find designing and making, architecture and engineering, art and science muddled up together so constantly and utterly that a once-and-for-all process of dissociation in an age of reason or enhanced technology appears implausible.' *Ars sine scientia nihil est* and *scientia sine ars nihil est*: partnership and recognition and appreciation of complementary skills are in the interests of all construction professionals.

US President Harry Truman said in 1948, 'You can achieve anything so long as you don't mind who takes the credit.' Pride in a quality product should outweigh personal recognition. However, it is galling when the credit is given to a quite different professional group. The media often credit scientists with having done something which is clearly the result of engineering input, and architects (we may feel) receive more than their share of credit for civil and structural engineering projects. Most projects are in fact partnerships between a number of parties who contribute different skills. Credit for success or blame for failure should usually be shared among a number of people or organizations. Perhaps it matters only insofar as aspiring members of the civil engineering profession are deterred by the severely subordinate role that civil engineers perceive themselves to hold. To complain continually of lack of recognition can rapidly be interpreted as paranoia. A gentle education of those who write in newspapers or broadcast on television or radio is required – they have an influence on public perceptions. The historical record is not always entirely objective.

Chapter 1
Materials of civil engineering

In searching for economical and efficient solutions to the problems that have been posed, engineers seek effective exploitation of the mechanical properties of the materials that are being used: properties such as the stretchiness or stiffness (the change in dimensions as the material is subjected to changes in load) and the strength (the limit to the amount of load that can be supported). Through experience and understanding, engineers have become more adept at exploiting these properties in order to produce increasingly daring structures. Effective exploitation implies that civil engineers are able either to control these properties to suit the application or to understand the nature of the properties of the materials with which they are presented. For example, the steel and concrete that we see in structures above the ground can be designed to have chosen stiffness and strength. These are manufactured materials which may have to be transported a long way to reach a construction site. The costs of transportation may be regarded as excessive, and locally sourced wood and rock may be available instead – their strength and stiffness are as you find them. Rock is just one constituent of the more general range of materials that make up the ground. The softer materials are designated as soils though the boundary between weak rocks and hard soils is not well defined.

Soil

All civil engineering constructions sit upon or sit in the ground. Tunnels pass through the ground; foundations for buildings and bridges are excavated from the ground; aeroplanes land on the ground; motor cars and railway trains drive on roads or rails laid on the ground; and ships berth against structures which are attached to the ground. So the behaviour of the soil or other materials that make up the ground in its natural state is rather important to engineers. However, although it can be guessed from exploratory probings and from knowledge of the local geological history, the exact nature of the ground can never be discovered before construction begins. By contrast, road embankments are formed of carefully prepared soils; and water-retaining dams may also be constructed from selected soils and rocks – these can be seen as 'designer soils'.

Today tourists flock to Pisa (Figure 1) not because of the fine Romanesque duomo but because of the unfortunate incident of the belfry (which has resulted in the area around the duomo being called the *campo dei miracoli*) – they come to see the leaning tower. Like the duomo, the tower is an example of Romanesque construction with semicircular arches round the lower storey and semicircular arcades of arches around each storey of the tower. The few window openings on the staircase which spirals round inside the wall are small. The bells are hung within the top storey. What went wrong?

Soils are formed of mineral particles packed together with surrounding voids – the particles can never pack perfectly. The mineral particles come from the erosion of rock: sands are formed of particles of quartz or feldspar formed by mechanical erosion; clay minerals are silicates and aluminates resulting from weathering of parent rocks under conditions of high temperature and humidity. It may be surprising that granite, which is

1. The tower of Pisa would not attract tourists if the foundation conditions had been more uniform. Recent removal of small quantities of clay from the foundation has stabilized the tilt

naturally a very strong structural material, weathers to form the clay mineral kaolinite, used for making porcelain. Porcelain, after it has been formed and baked at high temperature, is once again very strong and brittle.

Clay particles are small – officially smaller than 2 microns (2 millionths of a metre) in size – but soils can contain a very wide range of particle sizes depending on the source material, the manner of the weathering process, and the nature of the transportation process to the present location. Fine particles can be carried by wind; coarser particles can be carried by water; ice in the form of glaciers slowly moving down hillsides can transport huge boulders which may be subsequently discovered when the ground conditions at a site chosen for construction are being explored.

The voids around the soil particles are filled with either air or water or a mixture of the two. In northern climes the ground is saturated with water for much of the time. For deformation of the soil to occur, any change in volume must be accompanied by movement of water through and out of the voids. Clay particles are small, the surrounding voids are small, and movement of water through these voids is slow – the *permeability* is said to be low. If a new load, such as a bridge deck or a tall building, is to be constructed, the ground will want to react to the new loads. A clayey soil will be unable to react instantly because of the low permeability and, as a result, there will be delayed deformations as the water is squeezed out of the clay ground and the clay slowly *consolidates*. The consolidation of a thick clay layer may take centuries to approach completion.

The site chosen for the construction of the belfry at Pisa, begun in 1173, provided a rather poor foundation. The irregular layering of two different clays gave rise to a variable time-dependent deformation under the foundation of the tower with the result that it started to tilt only a few years after construction had begun.

The 12th and 13th centuries were unsettled times in Italy with the different city states of Pisa, Genoa, Lucca, and Florence all vying for dominance. Battles demanded money and people, and so the construction of the belfry took second place. When the masons under the direction of Giovanni di Simone resumed construction in 1272 they introduced a curve in the vertical alignment of the tower so that the courses of stone were once again being laid horizontally. This stratagem proved inadequate and the tower continued, slowly, to tilt further. Construction stopped again in 1284 and the belfry was only completed in 1372. The top storey very obviously makes another valiant attempt to establish a properly vertical line.

A leaning tower may be a valuable draw for tourists but as the tilt increases, the tower becomes progressively less stable. The tilted loading tries to encourage additional tilt of the foundation. There are many leaning towers in Italy: the skyline of Venice contains an impressive collection. There were once more. The brick-built campanile of San Marco in Venice collapsed in 1902 and was rebuilt. In 1989 the civic tower in Pavia collapsed without warning, killing three people. It was leaning but not dramatically: failure may have occurred because of variable movements in the cement mortar in which its bricks were set or because of slow, time-dependent changes in the properties of the bricks themselves rather than because of foundation problems. But the failure of one leaning tower naturally raised concerns about the safety of another one.

In 1990, careful monitoring of the Pisa campanile showed signs that the tilt of the tower was accelerating and there were signs of distress in stonework on the down-tilt side. Of the possible options, the only realistic one was to make some intervention in order to safeguard the tower in its current tilted position. Tampering with the foundation of a structure which is on the brink of collapse carries the risk that it will in fact trigger that collapse. This daring decision was taken by an international committee of geotechnical engineers (civil engineers with a particular interest in the behaviour of the ground). Modest safety measures were installed in the form of lead weights on the up-tilt side of the foundation together with cables which could have provided some slight restraint if the tilt had started to accelerate. Then small chunks of clay were extracted from under the up-tilt side of the foundation until the tower had rotated backwards to the inclination which it had had in about 1838. Its lifetime was thus successfully extended.

Timber

At Pisa and other cathedrals timber was used to construct temporary access platforms and supporting structures to enable the masonry building to be completed. Timber provides the internal structure for the *conventional* roof that is seen from outside the duomo. Provided there are trees around, timber has the obvious advantage over stone of being easier to work with. It obviously has the disadvantage of being combustible. Few timber buildings remain from the Middle Ages or earlier and, where they do, there is the constant risk of destruction. The Great Fire of London of 1666 started in a baker's shop and destroyed the old St Paul's Cathedral and some 13,000 houses. Nearly an entire block of the city of Trondheim, Norway, was destroyed in December 2002 by a fire which started in the kitchen of a restaurant. Also, timber has a tendency to rot if it becomes permanently damp and is eaten by bugs of various sorts and has a tendency to warp as it seasons and takes up or loses moisture.

On the other hand, trees grow in a way which is structurally efficient. The root systems respond to the external loadings applied by the wind in order to anchor a tree to the ground, whereas the branches are sufficiently flexible to move under wind and other environmental loadings. The cross section of a tree reduces from the roots upwards and outwards as the forces required to sustain or resist the movements of the more distant parts of the tree themselves become smaller. Wood can support compression loads very effectively and can support tensile loads, provided they are applied parallel to the grain (along the branches). Inspection of the wounds left by a branch or a trunk splitting will reveal the very fibrous nature of the growth and hint at the way in which these fibres might be strong individually but become readily *delaminated* if the applied loads try to pull them apart. The shape of a tree – a trunk with a single branch – can be deliberately exploited to provide a single curved *cruck* for a house or barn.

The great south gate of the Todaiji temple complex in Nara, Japan, (Figure 2) was built in the 12th century at about the same time as many of the great Gothic cathedrals of Europe. The decoration is elaborate and serves to conceal some of the structural interconnections, with horizontal tie beams slotted through the main supporting vertical posts (a *trabeated* style of building). The detail of the substructures devised to permit the eaves of the roof to overhang as far as possible and to enable a high roof ridge, apparently two storeys up, to be formed over a single storey internal space shows great ingenuity. This was the product of cultural exchanges between Buddhists in Japan and China led by the priest Shunjobo Chogen who brought Chinese craftsmen to work alongside Japanese carpenters. Japan is, of course, a

2. The great south gate (Nandaimon) of the Todaiji temple complex in Nara, Japan is typical of the use of 'trabeated' timber construction – beams and columns

country of frequent earthquakes: timber structures such as these old temple buildings have enough flexibility in their joints to be able to withstand significant horizontal shaking without damage.

Stone

'Hardwick Hall – more glass than wall' was built by Robert Smythson for Bess of Hardwick in the 16th century, on a hilltop, as a deliberate statement of wealth and power. The individual panes of glass were small and irregular, the source of that dappled charm of old buildings, but the internal result was of great light. The Romanesque cathedrals such as Pisa give a feeling of solidity and security. The columns in the nave of Durham cathedral, in northern England, are more than 2 metres in diameter and security or even defence was often needed in those politically unstable times. But within a century or two the cathedral masons across Europe had learnt from experience how to exploit the properties of their primary building material, stone, in order to produce structures in which light – apparently in defiance of solidity – seems to be the most significant feature (although at many of the cathedrals one might think of *colour* rather than *light* since the enormous windows are usually filled with magnificent mediaeval stained glass).

Rock (or stone) is a good construction material. Evidently there are different types of rock with different strengths and different abilities to resist the decay that is encouraged by sunshine, moisture, and frost, but rocks are generally strong, dimensionally stable materials: they do not shrink or twist with time. We might measure the strength of a type of rock in terms of the height of a column of that rock that will just cause the lowest layer of the rock to crush: on such a scale sandstone would have a strength of about 2 kilometres, good limestone about 4 kilometres. A solid pyramid 150 m high uses quite a small proportion of this available strength. If the rock can be easily cut then it can readily be used for the carving of shapes and figures (but is also more likely to weather and decay).

The builders of the Egyptian pyramids had the task of creating a massive rock structure around a small burial chamber. The small hole created local stress concentrations which they could relieve by leaning slabs of stronger stone against each other to form an internal pitched roof. Or alternatively they could create an internal protection by placing blocks of stone one on top of the other on each side of the void, projecting out (*corbelling* – see Figure 3) over the void gradually further and further until the void was closed by a stepped roof – much as one would if trying to create a bridge with children's bricks to span a gap wider than the longest brick.

The Romans developed the semicircular arch as a structural form. An arch formed of a series of shaped *voussoirs* (carefully prepared wedge-shaped stones) is extremely strong once it is complete, provided it is able to push sideways against unyielding abutments (Figure 4). During construction the incomplete arch has to be

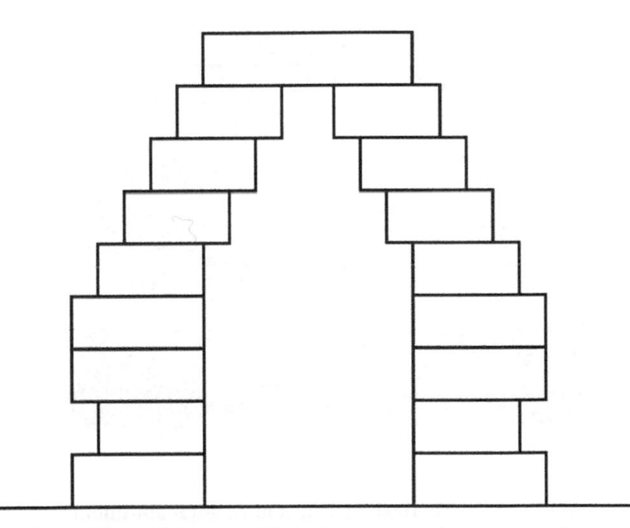

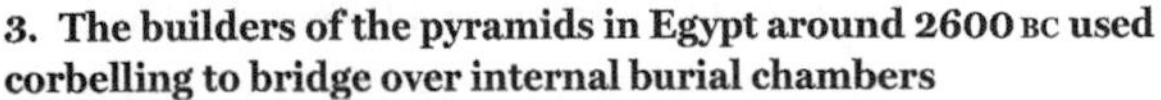

3. The builders of the pyramids in Egypt around 2600 BC used corbelling to bridge over internal burial chambers

supported using appropriate *centring*, usually made of wood (Figure 5). A little mortar on the radial joints between the voussoirs can provide some extra strength, but is not essential if the stones are carefully prepared. The centring is typically removed by pulling out carefully placed timber wedges. There is usually a little inevitable movement of the masonry as it takes up the load.

When the centring for the Pont de Neuilly over the Seine just west of Paris was going to be removed in 1772, the engineer, Perronet, made sure that the centring for each of the five spans was removed simultaneously with a great splash in order to distract attention from the movement of the bridge which the crowds of spectators (including Louis XV) might otherwise have found a little disturbing. If the abutments move outwards a little, as they usually will, then

4. An arch converts vertical loads to a horizontal abutment thrust. This is one of Andy Goldsworthy's 'striding arches' near Moniaive

5. Until the arch is complete it requires temporary support in the form of timber 'centring', shown here for Blackfriars Bridge under construction across the Thames (it was opened in 1769)

the arch will spread a little and cracks open between some of the voussoirs – see Figure 6. This is not necessarily a problem provided the loads applied to the arch (which might form the main load carrying element of a bridge or aqueduct) can find a route through the contact points of the voussoirs in whatever slightly rotated positions they now find themselves. Careful study of masonry structures, old and new, will usually reveal some modest (harmless) cracking from which the mechanism of displacement of the masonry can be deduced.

A continuous *barrel* vault over a space such as a hall, a wine cellar, or the nave of a church, is just a series of arches in a row, all pushing sideways uniformly and with solid supporting side walls resisting this push. However, if these side walls consist mostly of space (windows) then some other means of supporting the vertical load (the weight of the vault and the roof) and withstanding the

6. Many arches are able to carry loads even when cracks open between rotated voussoirs, as in this bridge near Constantine (Algeria)

sideways push has to be devised. So these forces have to be concentrated into individual columns between the windows. With everything being done to emphasize height, the columns supporting the vault to the nave of a church have to be slender. There is no difficulty in the columns carrying the vertical load from the vault and roof. Even in a cathedral like Chartres, near Paris, where the vault is 37 metres above the floor of the nave, the margin of safety appears to be large compared with the *strength* of the rock, which is of the order of a few kilometres. As we look up into the vault we can see the ribs channelling the loads into the columns. Unlike a bridge, there are no earth abutments at the top of the columns so a stone *abutment-substitute* has to be created. The horizontal push from the vault, and the loading from winds, can be absorbed by using elegant flying buttresses (Figure 7), which carry the load over the side aisles to lower external buttresses. Pinnacles on top of these buttresses push the load more effectively into the masonry.

For most Gothic buildings the general scheme or concept would be defined by the dean and chapter of the cathedral but they,

7. The arched vault of a Gothic cathedral such as Chartres requires support in the form of flying buttresses which carry the lateral loads (and wind load) over the aisles

having no expertise in construction, would expect the master mason (engineer) with his team of masons and labourers to devise the structural (and decorative) detail and to work out how to build it. Such teams were peripatetic, moving from one construction site to the next as money ran out or became available. They would learn new ideas for the next project and there would, especially at a time of widespread church building as in the 13th century, be a lot of exchange of skills and quite a rapid transfer of knowledge over large distances. We can now rationalize the detail of the structure of Gothic buildings and use more or less sophisticated methods of calculation to conclude that the masonry is still fit for purpose. The masons could only build on experience transmitted as geometrical rules of proportion which defined increasingly daring but empirically safe structural forms: the unsafe ones are no longer there for us to see. Serendipitously, but perhaps not surprisingly, some of the modern analyses can also be reduced to geometrical constructions.

Timber temporary structures enabled the masons to gain access to the working levels of the church and to provide support *centring* for the stone of the vaults and arches as they were being assembled. It would not have been practicable to erect scaffolding from the floor to the vault, and very difficult to arrange the safe removal of the centring to allow the vaults to take up their loads. Large churches are riddled with corner stairs and passages hidden in the larger columns and walls, giving access to temporary working platforms. The centring would have been strutted from projections and holes deliberately left in the walls below. Cranes (*engines*) driven by man-powered wheels (like large versions of hamster wheels, putting the energy to useful purpose) were used to lift blocks of stone.

While the builders of the Gothic cathedrals were becoming ever more daring (Figure 8) with their exploration of the possibilities of the stone arch, the Khmer builders of the temples of Angkor in Cambodia (Figure 9) were still, in the 13th century after five or six

8. The later Gothic flying buttresses at Prague reflect the increasing confidence and daring of the masons in their use of stone as a construction material

9. The temple builders in Angkor Thom, Cambodia, were still using corbelling to create archways in AD 1200

centuries of building, sticking firmly to a *trabeated* form of construction using horizontal lintels and vertical supporting posts. They had not been told about the arch and could only bridge spaces with corbelled structures as used over the inner chambers within the pyramids several thousand years earlier.

The knowledge possessed by the European masons (engineers) was kept within the group as far as possible: they had to train their successors but had also to give the impression that their largely empirical knowledge, based on geometrical rules whose prior success had been observed, was too complex for the layman, or amateur, or ecclesiastical client to grasp. Where failures occurred they were usually *geometrical*: extremely slender and tall columns became susceptible to instability through buckling if there was some slight change in alignment as a result of foundation settlement. The emergence of theoretical ideas which might support their designs (or might not) was regarded as a threat in somehow removing this aura of secrecy and skill and potentially opening their profession to anybody of mathematical education.

Iron and steel

Iron has been used for several millennia for elements such as bars and chain links which might be used in conjunction with other structural materials, particularly stone. Stone is very strong when compressed, or pushed, but not so strong in tension: when it is pulled cracks may open up. The provision of iron links between adjacent stone blocks can help to provide some tensile strength. The concept was straightforward but the technology was challenging. The Industrial Revolution, in the 18th century, resulted in the development of techniques for the reliable production of iron (and later steel), which were then put to use in the creation of more exciting structures than could be produced with masonry alone.

Cast iron can be formed into many different shapes and is resistant to rust but is brittle – when it breaks it loses all its strength very suddenly. Wrought iron, a mixture of iron with a low proportion of carbon, is more ductile – it can be stretched without losing all its strength – and can be beaten or rolled (*wrought*) into simple shapes.

Steel is a mixture of iron with a higher proportion of carbon than wrought iron and with other elements (such as manganese or titanium or chromium) which provide particular mechanical benefits. Mild steel has a remarkable ductility – a tolerance of being stretched – which results from its chemical composition and which allows it to be rolled into sheets or extruded into chosen shapes without losing its strength and stiffness. There are limits on the ratio of the quantities of carbon and other elements to that of the iron itself in order to maintain these desirable properties for the mixture. Its development was thus associated with both the provision of close control of the chemical content in the blast furnaces (then the Bessemer converter, and latterly electric arc furnaces) and the construction of powerful rolling mills.

There are many iconic photographs of the high-riggers on skyscrapers in New York, showing a disregard for safety precautions

10. Site safety rules would today frown upon the antics of these workmen assembling the steel frame of a skyscraper in New York. However, the I cross-section of the beams and columns is apparent. The web carries the transverse load and the flanges resist bending – and improve resistance to buckling

that would not be accepted today (Figure 10). If you look at a steel-framed building while it is being constructed you will see that most of the steel elements, vertical or horizontal, have an H section consisting of two parallel plates (*flanges*) connected by another plate (*web*). Steel is very strong and stiff in tension or pulling: steel wire and steel cables are obviously very well suited for hanging loads. If you push down on a long thin strip it will not remain straight but will *buckle* sideways, undergoing geometrical failure. If you are lucky, it will return to its original shape when you stop squashing it. This is the geometrical failure towards which the increasingly slender and tall stone columns of Gothic cathedrals were heading.

The buckling of a thin steel section might seem to indicate that the material has a low stiffness when it is compressed. This geometric loss of stiffness can be improved in two ways. If we hold the centre of the thin strip to stop it moving sideways and then push down we will find that the load at which the buckling instability occurs is about four times larger than for the unrestrained strip (Figure 11). The mode of buckling now has two bows, one each side of the central restraint. So, shortening the *unsupported* length is one way of improving the ability of a structural member to support compression.

The other way is to make it harder for the structural member to bow. A thin plate on its own has very little bending resistance. If the plate is incorporated into an H section then the bending becomes more difficult. However, even an H section steel girder will be much stronger when being pulled than when being compressed. In compression, buckling will eventually occur. The propensity to buckle thus depends on the cross-section of the girder and also on its length.

The design of the river crossing of the Firth of Forth (Figure 12), heading north from Edinburgh, was subjected to great scrutiny and conservatism of design because of the failure, in 1879,

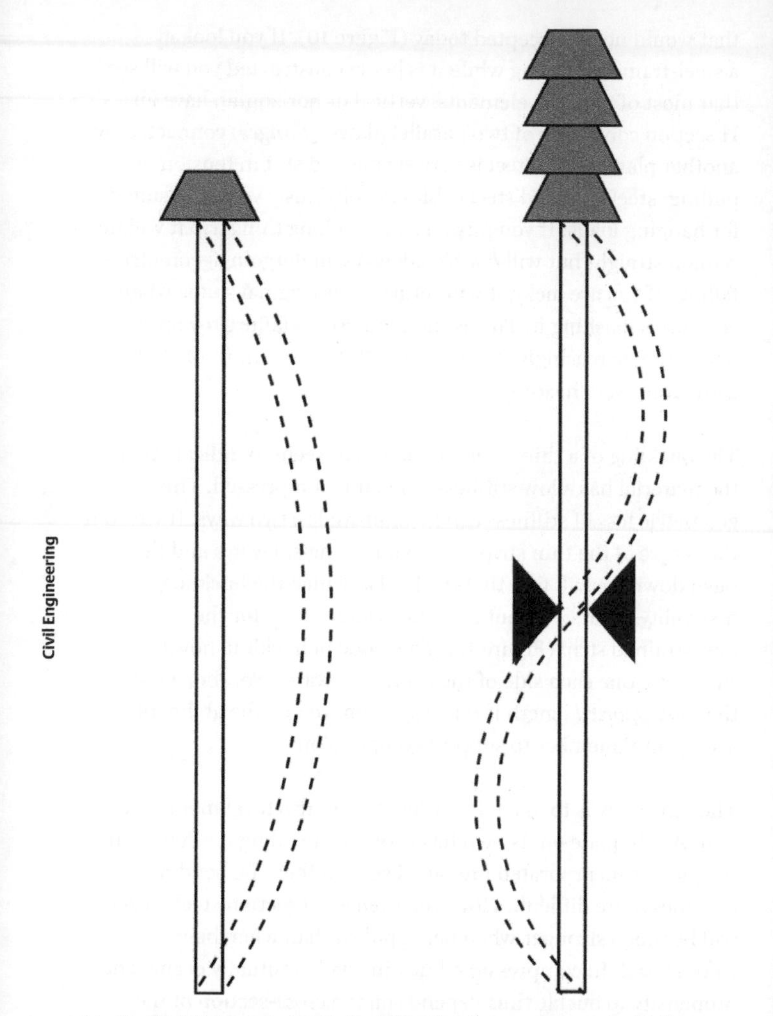

11. (Left) A thin strip, loaded axially, develops a *geometrical* failure called *buckling*. (Right) If the strip is prevented from moving sideways at its midpoint then the load it can carry before it buckles is increased fourfold

12. The Forth Railway Bridge, near Edinburgh, was designed to resist much higher wind loading than the Tay bridge, some 50 km further north, which collapsed in a storm in 1879

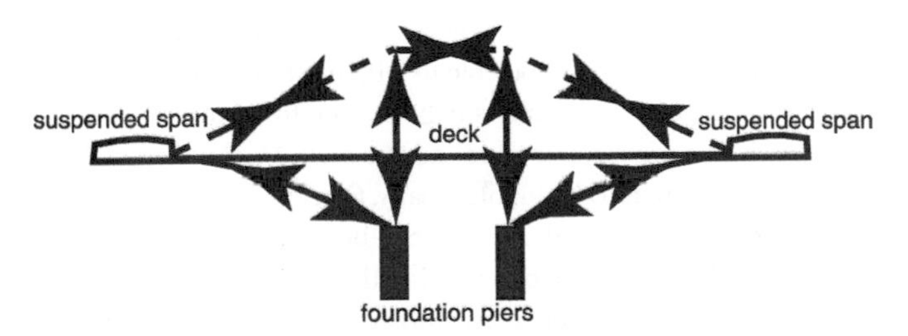

13. The topmost members of the main truss of the Forth Railway Bridge (dotted) act in tension (pulling); the vertical and lower members of the truss (solid) act in compression (pushing). The truss is triangular in cross-section, giving extra stability. The suspended spans are, in comparison, very lightweight

of the railway bridge across the River Tay – the next estuary to the north. The designer for the Forth Bridge Company, which was providing a vital link between sections of the railway network of the North British Railway Company, was Sir Benjamin Baker with Sir John Fowler. The result was the iconic structure, completed in 1890, somehow typifying Victorian engineering, and it is still in regular use for all railway traffic heading from Edinburgh to Dundee and Aberdeen. Such icons are not necessarily things of beauty but this one shows very well how it carries the loads (Figure 13), with strongly braced three-dimensional frames sitting over each pier separated by lighter structural sections which balance between the strong frames. It was the first major structure in Britain to be constructed entirely from steel (the contemporary Eiffel Tower in Paris was constructed of wrought iron). A similar structural arrangement was used for the Canadian National Railway bridge across the St Lawrence river, upstream of Quebec, completed in 1919.

Masonry (stone) arches have been recognized for more than two millennia as providing a strong way of bridging a gap in order to support loads of people or traffic. The arch converts the vertical forces provided by the steady or varying loads into horizontal forces at the abutments, each side of the gap. The abutments need to be sufficiently strong to be able to withstand this horizontal push. The shape of the arch has to be chosen to ensure that the masonry is in compression all round the arch. Masonry is heavy and may not be readily available locally. Steel provides a much lighter structural material. An arch can be formed out of a steel lattice with the cross-bracing ensuring that the unrestrained lengths of the compression elements are short enough to eliminate the possibility of buckling.

The Sydney Harbour Bridge (Figure 14), which was completed in 1932 with a main span of 503 m, is just such a steel arch bridge, but with the deck running through the arch at half height. It is very similar in appearance and structural form

14. The deck of the Sydney Harbour Bridge is carried on a steel arch with plenty of vertical and diagonal bracing to ensure that the arch will not buckle

to the less well known Hell Gate Bridge in New York, a railway bridge 310 m long, designed by Gustav Lindenthal and completed in 1916. The deck is partly hung from and partly supported on (towards the ends of the main span) the steel arch which is terminated with a masonry tower at each end which serves the same structural purpose as a pinnacle over the outer wall of a Gothic cathedral, where it receives the force from a flying buttress. Its weight helps to push the resultant abutment force from the arch more directly downwards into the foundation. Aesthetically the towers give a reassuring sense of solidity to the bridge. The upper steel members, which also seem to form an arch, do not actually reach the masonry towers. They provide stiffening and lateral support to the arch itself in order to prevent it from buckling.

Concrete

Concrete is a sort of artificial rock (conglomerate) which can be cast to almost any chosen shape and size. The Romans discovered that mixing pozzolana – a volcanic ash product found in the region around Vesuvius – with lime and water produced a reaction which, when complete, left a very strong rock-like material. The dome of the Pantheon in Rome was built in around AD 126 using this early concrete. A large quantity of massed concrete around the lower levels of the dome was used to absorb the horizontal thrusts. The Pantheon is intended to be admired from within (the coffered cut-outs from the underside of the dome help to reduce its weight) rather than from without. It remained the dome with the largest span in the world until the dome of Santa Maria del Fiore in Florence was completed.

For some applications, all that is required is the mass of the concrete, and its necessary mechanical properties are equivalent to those of an artificial rock. For example, with appropriate ground conditions, concrete gravity dams provide a straightforward means of blocking a valley and retaining a reservoir.

The Colorado River runs from the Rockies in Colorado and Wyoming through Utah, Arizona, and Nevada to the Mexican border and the Gulf of California. Its name comes from the colour of the sediments that it carries down to the sea. But for river water to reach the sea (in another country) seems a waste when it could be used to support growing city populations. The Hoover Dam (originally called the Boulder Dam, from the canyon that it closes) approved by Congress in 1928, together with the 400 km aqueduct to Los Angeles, was completed in 1935, providing a major source of construction employment during the Depression years. Congressional approval was notionally dependent on the completion of an agreement between at least six of the seven states which had an interest in the Colorado water. But the agreed water distribution was based on flow rates which were greatly overoptimistic. The dam retains a water depth of about 180 m and retains a reservoir which is some 180 km long with a volume of about 35 cubic kilometres. It regulates the flow in the downstream river to allow abstraction for the aqueduct but also relies on this flow through its turbines to generate 2.1 GW of electricity, of which a sizeable proportion is required to power the five pumping stations that take the water over the hills to the Californian coast.

A dam is just a big plug in a valley. This plug can be made in many different ways. Childhood experience of damming streams shows that compacted earth will restrict the flow and allow a (modest) lake to form. Including some stones or small rocks helps to give some extra solidity. But water is quite good at finding chinks – through or under or round the sides of the dam. Even if these holes are blocked, when the lake fills to the top of the dam the overflowing water will erode weaker elements of the downstream face. Few of those youthful constructions survive for long.

The same problems face the designer of a real dam. Initially, small dams across the valley divert the river into pipes or tunnels so that the subsequent serious construction work can proceed in the dry. At the Hoover Dam parts of these diversion tunnels were incorporated into the eventual spillway system allowing controlled flow of water into the downstream river and also a safe means of discharge, without erosion of the dam, when the reservoir is full.

The pressure exerted by 180 m of water is high and has somehow to be transmitted to the base and sides of the gorge in which the dam is to be built. An arch is an effective way of transmitting vertical loads from bridge traffic to the horizontally resistant abutments. An arch dam performs the same mechanical process but turned through 90 degrees. The Hoover Dam is a combined gravity arch concrete dam. The convex arched upstream section in plan takes the water pressure to the sides of the gorge. The descriptor *gravity* refers to the sheer bulk of the dam sitting on the floor of the gorge and resisting downstream movement. The dam is 200 m thick at its base and 14 m thick at the crest, which is 380 m long, and it has a volume of 2.5 million cubic metres.

As concrete sets, the chemical reactions that turn a sloppy mixture of cement and water and stones into a rock-like solid produce a lot of heat. If a large volume of concrete is poured without any special precautions then, as it cools down, having solidified, it will shrink and crack. The Hoover Dam was built as a series of separate concrete columns of limited dimension through which pipes carrying cooling water were passed in order to control the temperature rise. Evidently the junctions between the blocks needed to be sealed and the cooling pipes to be filled, and the rock beneath and beside the dam to be rendered impermeable.

Grout is a fine mixture of cement, water, and sand which can be pumped into gaps and fissures. Holes drilled in the valley floor were used to inject a *grout curtain* in order to seal the rock to a depth sufficient to limit any leakage from the full reservoir. Similarly, having understood the nature of the geological structures present in the sides of the valley, these rocks were grouted both to prevent flow of water and also to strengthen them to resist the abutment forces from the dam.

The importance of understanding the geology of a site for the construction of a dam was emphasized by the failure in 1959 of the Malpasset Dam in south-east France. This was a double curvature mass concrete arch dam – curved both horizontally (like Hoover) but also vertically. It was the thinnest arch dam ever built – 7 m thick at the base and 1.5 m thick at the crest – and about 60 m high and 190 m between abutments. Completion in 1954 was followed by a series of dry years (and a problem with a recalcitrant landowner) and it was not until the extremely heavy rains in late 1959 that the reservoir filled very rapidly. The dam failed on the evening of 2 December killing 423 people downstream. Protracted investigations revealed weak seams in the rock of the left abutment which had displaced slightly under the rapidly rising water load. This displacement was enough to destroy the elegant structure.

Reinforced and prestressed concrete

Concrete is mixed as a heavy fluid with no strength until it starts to set. Embedding bars of a material such as steel, which is strong in tension, in the fluid concrete gives some tensile strength. *Reinforced concrete* is used today for huge amounts of construction throughout the world. When the amount of steel present in the concrete is substantial, additives are used to encourage the fresh concrete to flow through intricate spaces and form a good bond with the steel.

For the steel to start to resist tensile loads it has to stretch a little; if the concrete around the steel also stretches it may crack. The concrete has little reliable tensile strength and is intended to protect the steel. The concrete can be used more efficiently if the steel reinforcement, in the form of cables or rods, is tensioned, either before the concrete has set or after the concrete has set but before it starts to carry its eventual live loads. The concrete is forced into compression by the stretched steel. The working loads that the concrete has to support would have been expected to generate tension in the concrete. As a result of the prestressing, these loads are carried by reducing this initial compression of the concrete.

Such *prestressed concrete* gives amazing possibilities for very slender and daring structures (Figure 15). The analysis of such structures provides its own challenges: the concrete must be able to withstand the tension in the steel, whether or not the full working loads are being applied. For an arch bridge made from prestressed concrete, the prestress from the steel cables tries to lift up the concrete and reduce the span whereas the traffic loads on the bridge are trying to push it down and increase the span. The location and amount of the prestress has to be chosen to provide the optimum use of the available strength under all possible load combinations. The pressure vessels used to contain the central reactor of a nuclear power station provide a typical example of the application of prestressed concrete.

All nuclear power generation so far relies on the principle of nuclear fission: when bombarded with neutrons, uranium

15. Prestressed concrete provides possibilities for extremely slender structures such as this ribbon footbridge at Shirakawa in Japan

235 decays into other elements (such as krypton and barium) with the release of several neutrons. These neutrons are available to bombard other uranium atoms to encourage further degradation, each time with the release of energy. At the centre of a nuclear power station is a reactor in which this process occurs in a controlled fashion. Control is provided by a combination of nuclear moderator and control rods made of material which is 'poisonous' to neutrons and thus can absorb them or slow them down so that their thermal energy can be extracted. Graphite is a good moderator, and the cores of nuclear reactors are often formed of carefully engineered graphite blocks. The coolant fluid used to extract heat from the core can also have moderating properties.

The nuclear power station at Torness in south-east Scotland (Figure 16), using advanced gas-cooled reactors, was among the last of this type to be built in the UK at the end of a programme of nuclear investment which began in the 1950s. It has a generating capacity of 1.36 GW and came on power in 1988 after an 8 year construction period. Its operating life is presently destined to end in 2023. The two reactors use carbon dioxide at 40 times atmospheric pressure as the coolant of the radioactive core. Water is pumped through a heat exchanger in the reactor where it absorbs heat from the circulating carbon dioxide and is turned to steam at a temperature similar to that used in typical coal-fired power stations. The steam expands through the turbines and in generating power loses pressure and partially condenses to water. The condensation process is completed in a cooling loop, and the resulting liquid is ready to start the cycle once more. The condensation process requires cooling water, and typically power stations are located by a convenient river or other source of water – Torness is on the coast.

There are many civil engineering contributions required in the several elements of the power station – and certainly many interactions with specialists from other branches of engineering. The electricity generation side of a nuclear power station is

16. The grey cuboids of the power station at Torness, Scotland, give no clue of the prestressed concrete nuclear reactors and power generation plant to be found inside

subject to exactly the same design constraints as any other power station. Pipework leading the steam and water through the plant has to be able to cope with severe temperature variations, rotating machinery requires foundations which not only have to be precisely aligned but also have to be able to tolerate the high frequency vibrations arising from the rotations. Residual small out-of-balance forces, transmitted to the foundation continuously over long periods, could degrade the stiffness of the ground. Every system has its resonant frequency at which applied cyclic loads will tend to be amplified, possibly uncontrollably, unless prevented by the damping properties of the foundation materials. Even if the rotating machinery is being operated well away from any resonant frequency under normal conditions, there will be start-up periods in which the frequency sweeps up from stationery, zero frequency, and so an undesirable resonance may be triggered on the way – just as there is often a slightly alarming stage as a washing machine moves into the spin stage.

The coolant under high pressure in the nuclear reactor has to be retained in a pressure vessel: complete integrity of the containment vessel is important. Torness has a central gas-tight liner formed of 13 mm thick carbon steel and around this a prestressed concrete pressure vessel 5.5 m thick which acts as a biological shield. A vessel under internal pressure is trying to pull itself apart in all directions. The concrete in the pressure vessel must be kept under compression by embedding a dense array of steel wires which are pretensioned before the internal pressure is applied. The stretching of the wires is equilibrated by compression of the concrete. All the internal pressure now does is to reduce this compression.

Plastic

The use of plastic or composite materials (other than reinforced concrete) in conventional civil engineering structures has not developed very rapidly. Such plastic composites tend to be more expensive than their metallic equivalents so, unless weight is of particular concern (as in aircraft structures), there is no natural economic driver for their use. Carbon fibre sheets have been used to repair or strengthen existing structures to improve their resistance to earthquakes, where their light weight makes them easier to handle and position. Construction companies having been found to be reluctant to blaze a trail in the use of an unfamiliar material, the all-plastic footbridge at Aberfeldy, Scotland (Figure 17) was erected by students (without any need for high capacity lifting 'engines').

To link the Airbus A380 (Figure 18) and civil engineering may seem fanciful. However, an aircraft is a large structure, and the structural design is subject to the same laws of equilibrium and material behaviour as any structure which is destined never to leave the ground. The aerodynamic loads on the wings are of course an essential part of the loading for an aircraft; for most land-based structures the wind loading will be a lesser component

17. Civil engineering contractors have hesitated to tackle new construction materials. The footbridge at Aberfeldy was built entirely from plastic by Dundee University students under the direction of Bill Harvey

of loading. The A380 is an enormous structure, some 25 m high, 73 m long and with a wingspan of about 80 m; the area of the wings is about 850 m^2; and the aircraft can carry some 853 passengers, on two floors, if fitted out only for economy class. For

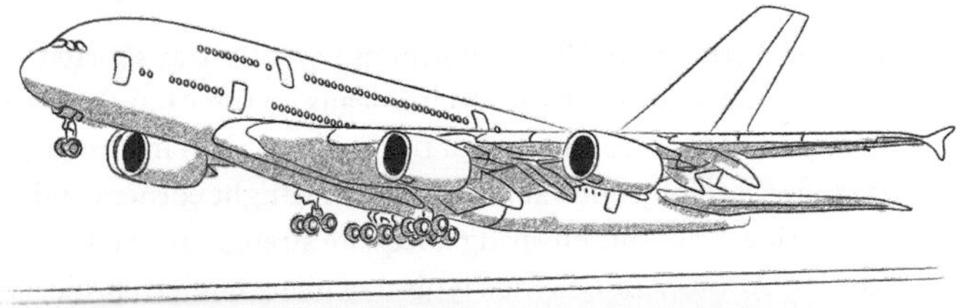

18. Any aeroplane is subject to the same laws of structural mechanics as conventional ground-based structures. The Airbus A380 has dimensions which are similar to those of a large building. Use of plastics in aircraft structures provides strength with reduced weight

comparison, St Paul's Cathedral in London is 73 m wide at the transept; and the top of the inner dome, visible from inside the cathedral, is about 65 m above the floor of the nave.

The structure of the A380 makes much use of plastic – composite – materials in its quest for lighter, stronger structures. Such composite materials obtain their desirable strength and stiffness properties from fibres of glass or carbon which are carefully aligned – exactly like the steel in reinforced concrete – in such a way as to best resist the stresses developed by the loadings applied to the structure – especially those applied to the wings. These strong fibres are protected by a weaker and less stiff matrix. The resulting materials are lighter than most metallic materials of the same or greater stiffness and strength.

But the Airbus A380 is also a wonderful example of the elaborate management and organization required for any complex engineering project. The aircraft are ultimately fitted out in Hamburg having been previously assembled in Toulouse. The rear sections of the fuselage come from Hamburg; the wings are manufactured in the UK; some of the front sections of the fuselage are manufactured in St Nazaire, France; the belly and tail of the plane come from

Cadiz in southern Spain. These enormous parts are transported from their various origins by sea to Bordeaux on the Atlantic coast of France and then to Toulouse first by canal and then in a road convoy along carefully chosen routes, avoiding tight corners and narrow carriageways but ensuring adequate strength of the road surface. The Airbus A380 is thus typical of many modern projects in its need for the successful integration of a wide range of engineering skills and disciplines.

Road materials

'Before the Roman came to Rye or out to Severn strode, the rolling English drunkard made the rolling English road.' But after the Romans had left, the English reverted, and created lanes and roads that wound round the natural features of the landscape, skirted fields, linked farms with mills and with market towns – straight lines were the coincidental exception rather than the norm. The Romans had very clear notions concerning road layout – there was a wide paved carriageway with clear areas each side of the carriageway to remove the possibility of cover for ambush. These roads were clearly provided for military purposes. The concept of formal road building also disappeared with the departure of the Romans. The hope was that human and animal feet and cart wheels would compact the ground sufficiently to give a modest stability. But come the rains, and the track would once again become a mire. The farmer taking produce to and from market with a beast of burden could pick his way across the terrain, but armies with their chariots and military equipment (engines) needed something more serious to aid their rapid deployment. (One might imagine that elephants could be used to compact a layer of soil being placed for a road. Unfortunately, an elephant having once traversed a certain route will remember where it has been and always place its feet in the same spots: pachydermic compaction of an area of soil is not effective.)

Road design makes use of *designer soils*: materials from the ground are modified in order to give particular mechanical properties. There are also special materials that are used for construction of the road *pavement* (the surface on which the traffic runs). The principles of design of roads have not changed much over two millennia, even if the loading provided by the juggernaut lorries of today considerably exceeds that provided by Roman chariots and soldiers. The materials available for pavement construction have been improved, and the machinery available for placing and compacting those materials has developed – particularly in scale. The Via Appia which leads from Rome to Brindisi on the Adriatic coast, a distance of 360 Roman miles (about 530 km), built over a period of about 60 years around 300 BC, still exists. Traffic requires a strong running surface, resistant to wear. This layer needs some underlying support to prevent it spreading, and an underlying drainage layer to draw rainwater into lateral ditches to prevent the formation of a morass in wet weather. The Via Appia had a running surface some 4.1 m wide (to permit two-way military traffic) formed of shaped tightly fitting smooth stone slabs laid on a layer of gravel to provide the drainage: a labour intensive operation for which the availability of slave labour is helpful.

Adaptations were needed if the road had to cross swampy ground. The Romans developed a technique using osiers – willow branches – spread across the soft ground both to help distribute the load and also to tie together the road base material in order to discourage it from flowing out sideways. Today road designers use synthetic reinforcing materials – so-called *geogrids* or *geomembranes* – to achieve the same result.

When the features of good road construction were rediscovered in the 18th century by, among others, Thomas Telford (1757–1834) in Britain, efforts were concentrated on finding ways of producing a tightly packed running surface. John McAdam (1756–1836) became obsessed with the size of stones to be used.

He specified a maximum size of 20 mm for the stones: a road-building labourer could check whether a rock had been adequately broken by seeing whether the resulting stones could be fitted into his mouth. This size was chosen to be significantly smaller than the typical wheel rim thickness (100 mm). McAdam reckoned that the severity of Telford's camber was not necessary to guarantee sufficient runoff of rainwater, so his designs were more economical.

The material which we see so often on modern road surfaces, and known as tarmac (adam) or asphalt, preserving his name at least in part, was introduced in the early 20th century. Binding together the surface layers of stones with bitumen or tar gave the running surface a better strength. Tar is a viscous material which deforms with time under load; ruts may form, particularly in hot weather. Special treatments can be used for the asphalt to reduce the surface noise made by tyres; porous asphalt can encourage drainage. On the other hand, a running surface that is more resistant to traffic loading can be provided with a concrete slab reinforced with a crisscross steel mesh to maintain its integrity between deliberately inserted construction joints, so that any cracking resulting from seasonal thermal contraction occurs at locations chosen by the engineer rather than randomly across the concrete slab. The initial costs of concrete road surfaces are higher than the asphalt alternatives but the full-life costs may be lower. The eventual process of replacement of a concrete road surface may be more complex and disruptive to traffic flow than the repair of an asphalt surface and the driving experience, registering each construction joint in turn, may be less comfortable: there are various factors which might influence the choice.

While an aeroplane is not strictly a product of civil engineering, much of the infrastructure of an airport certainly is. There are buildings of greater or lesser complexity and architectural pretension, there are transport connections both between parts of the airport and between the airport and the outside world, and there are the runways. A runway pavement provides a foundation for a structure (the aeroplane) which can be placed anywhere within its area and which will generate dynamic impact loading as well. A runway is essentially just a short road and is subject to the same design considerations. Just as for heavy lorries, the load of an aeroplane is spread over a number of tyres in order to keep the individual loads to a satisfactory minimum. The runway spreads the load from these tyres into the underlying ground without significant deformation, temporary or permanent, of the runway surface. As with roads, runways may be surfaced with asphaltic or concrete materials. In choosing a material the likely need for, and frequency of, maintenance and repair will be particularly important since closure of a runway for more than a night-time slot will have major repercussions for the operation of the airport.

In many cases the engineer will need to make a choice between different candidate materials. There may be some uncostable influences on this choice – for example, aesthetic considerations – but it is likely that the choice will ultimately be governed by economic constraints. Choice of material should take account of the real costs involved. Different candidate materials will have different initial costs and different lifetimes: perhaps the cheaper material will need to be replaced much sooner. Conscientious full-life costing of civil engineering projects is important at a time when we are particularly concerned to make our activity on Earth more sustainable. The full costing should consider the energy requirements for the manufacture and transport of the materials that are being used (which may not be fully reflected in market prices) and also the costs of demolition and disposal or reuse of the construction materials. Many buildings are being constructed with an anticipated life of 25–50 years (compare the value for money of the longevity of the mediaeval cathedrals) – perhaps because it is expected that technological developments within that period will render current engineering solutions inappropriate.

Chapter 2
Water and waste

Water is not a construction material but it is of huge importance to mankind, both in the damage that it can do when uncontrolled and in the dependence that we have on its abundant provision. Water and waste illustrate the intricate interdependencies of civilized life. One man's effluent is another man's water supply. It is much easier to be casual about outputs until the effect on others becomes overwhelming.

A supply of fresh water is one of the essential elements of infrastructure which we often take for granted in towns or cities. Dwellings in remote rural areas may have their own wells or springs which are at the mercy of the weather. Carrying our own water when trekking into remote areas makes us all the more appreciative of the reliable tap back home. Child mortality in the developing world is dramatically reduced when the quality of the water available for drinking and cooking is improved. Failure of waste disposal systems following natural disasters – either directly destroyed or implicitly absent in temporary camps for displaced persons – leads to cholera and typhoid epidemics through contamination.

Flying over the western United States or central and western Australia, it is very obvious that the desert is nature's preference for much of the terrain. Settlements exist only under sufferance and these are certainly dependent on a supply of water from somewhere else, either below or distant from the site. Pumping water from deep wells or boreholes intercepting permeable aquifers uses *old* water which has probably spent many thousands of years percolating through the ground in order to reach the *aquifer* (the geological layer in which the water is stored and through which the water is able to flow). It is evidently quite easy for the rate of pumping to exceed the rate of replenishment, so that the water supply becomes a diminishing resource. Reliance on pumped water to supply an expanding urban area can be precarious. Water collected from rainfall in surface reservoirs or abstracted from rivers can be seen as *young* water: there is a more evident relationship between quantities of water that are available and demanded.

Rivers add their own problems to the water supply. Water abstracted upstream of a town will no doubt be discharged as waste water to be used as input for other towns further downstream. River water may be used many times before it is eventually allowed to find freedom in the sea.

Water

Water has been such an essential supporting component of civilization over the millennia that its availability has often governed the location of castles and villages. Equally, the provision of water and handling of waste for towns and cities has often required impressive civil engineering.

The island of Samos in the Aegean has a fortified town with the same name on its southern coast. Any fortified castle or town needs a supply of water (and food) in order to be able to survive a prolonged siege. A plentiful spring, separated from the town by the limestone Mount Ampelos (228 m high), provided a possible source. Transporting the water by contour channel would have been vulnerable to attack. The alternative, to tunnel through the mountain for a distance of about 1 km, is described by Herodotus, writing about 60 years after its completion in around 525 BC, as one of the greatest engineering feats of the Greek world. He actually names its designer/engineer (he uses the term architect, 'master builder') as Eupalinos.

Civil engineers are required not only to design their works but also put them in the intended locations. Skills of surveying are today greatly assisted by precision optical instruments, laser levels and global positioning systems feeding on satellites that are constantly orbiting the earth. Such modern devices may seem a bit like cheating to those surveyors who had to toil up mountains with heavy theodolites, hoping for conditions clear enough to be able to see their colleagues on adjacent peaks, all the time creating invisible triangles across the countryside which would be brought down to earth at a baselength which had been very precisely measured using chains, or rods of metal or glass. The precision that was achieved by such techniques in the 18th century was remarkable.

Eupalinos had no serious optical instruments, but he must have used some system of sighting beacons to establish the line of the tunnel under the mountain, and some sort of levelling device to establish the relative levels of the two entrances to the tunnel, constructed simultaneously from both ends. In the event the two tunnel headings met with a separation of 1 m vertically and 6 m horizontally. The water supply was carried in an earthenware channel at the bottom of a slot about 80 cm thick cut from the floor of the tunnel to depths up to 8 m, to ensure the necessary gradient of about 0.6 per cent providing a flow of some 1200 cubic metres per day, sufficient to supply a town of 20,000 inhabitants together with a fleet of ships, an army, and the construction workers brought in for the tunnel itself and other works of fortification. The channel itself was roofed over and the slot backfilled with rock debris. The spring at the north end and the supply reservoir at the city end are linked into the tunnel by buried conduits. The tunnel, 1.8 m × 1.8 m in section, doubled as a subterranean escape route from the fortified town – like the modern tunnel in Kuala Lumpur which doubles as a highway and a flood relief channel.

Travelling today on the Via Appia, or on other roads heading for Rome, you can see extensive remains of aqueducts conveying water to the city built between 312 BC and AD 226. There are other more spectacular remains of Roman water supply networks. The Pont du Gard, in southern France, supplied some 40,000 cubic metres of water per day to the city of Nîmes from a spring some 50 km away: the water took more than a day for its journey from spring to city. The Romans used lead or earthenware pipes when necessary to cross rivers by forming inverted siphons, but they preferred to use open channels as far as possible. The surveying feats in setting out the necessary low gradients for the flow over tens of kilometres were impressive.

Minerals of value, such as gold, are found where the geology is appropriate but not necessarily where there is water available for human or industrial consumption. The gold rush in Western Australia was driven by the same human weakness for the yellow metal as that in North America. The discovery of gold at Kalgoorlie in the 1890s, led ultimately to the creation of one of the largest open-pit mines in the world, but also triggered a need for water. Charles Yelverton O'Connor was appointed as chief engineer for Western Australia in 1892 and, having built the harbour at Fremantle, turned his attention to the problem of water supply for Kalgoorlie. His solution was to build a mass concrete dam in the hills just to the east of Perth and then construct a 0.75 m diameter pipeline for some 500 km through which to pump the water, with a series of eight pumping stations. The project took about five years to complete and water started flowing through the pipeline in 1903. Sadly, the constant sceptical criticism of the project and false accusations of corrupt practices to which he was subjected led O'Connor to commit suicide in 1902 by riding his horse into the sea. He is now regarded as a local hero.

Waste

A good supply of fresh water is one essential element of civilized infrastructure; some control of the waste water from houses and industries is another. The two are, of course, not completely independent since one of the desirable requirements of a source of fresh water is that it should not have been contaminated with waste before it reaches its destination of consumption: hence the preference for long aqueducts or pipelines starting from natural springs, rather than taking water from rivers which were probably already contaminated by upstream conurbations. It is curious how often in history this lesson has had to be relearnt.

At Mohenjodaro on the Indus River in what is now Pakistan from 2500–1500 BC there was an extensive network of drainage channels beneath the streets, lined with fired bricks and leading to soakaways beyond the urban area. The Minoans in Crete provided a similarly elaborate flushed drainage system beneath the palace at Knossos at around the same period. However, the Greeks did not spot the advantage of such a drainage scheme, and waste was thrown in the streets in a way that continued in many European cities into the 18th century.

The inhabitants of Rome did rather better. Large underground channels were constructed to drain low-lying areas of the city and channel rainwater from the streets. Lesser drains carried waste from public latrines, flushed by water from the public bath-houses, efficiently using grey water for this purpose, a concept that has been adopted in the past century in areas where water is in short supply (and which could rationally be adopted more generally). However, such municipal engineering for the public benefit did not extend far outside the Imperial City.

Visitors to the ruins of the town of Pompeii (240 km from Rome), destroyed by the eruption of Vesuvius in AD 79, will recall the paved main streets with raised pavements and occasional stepping stones. The spacing of the stones allowed for the spacing of the chariot wheels; while the gaps between and the height of the kerbs provided space for rainwater to flow and in the process clear some of the animal and human waste and other detritus from the streets. Pompeii did have a water supply to the public baths and street-corner fountains but did not have any organized subterranean system of sewage collection and disposal. Some houses had latrines, usually located in diminutive kitchens so that cooking waste could join everything else in the cesspit.

London embankment sewer

In the 19th century, the Industrial Revolution had resulted in a huge expansion of the city of London, but its estuarine location meant that gradients were shallow and there were no mountain torrents to arrive fresh from the springs, either to provide a supply of pure water or to provide a means of clearing the streets. From the early 17th century water had been brought by a New River from a spring some 65 km north of London but that was exceptional – over the next two centuries several companies were formed to sell water which had been drawn from the Thames itself; the further downstream these points of abstraction the more polluted the water became. There were many who publicly proclaimed the quality of the Thames water and complacently asserted that London was one of the healthiest cities on earth. Perhaps that was true as a relative comparison but certainly not as an absolute one. All the drains of London poured their filth into the Thames when the tide was out – otherwise the drains were below river level – and this polluted water was then washed up river by the rising tide. Cesspits for houses and tenements were being sunk so deep that they intercepted the aquifer from which many of the wells of London took their water.

There was also a strong reluctance to admit the possibility that disease could be transmitted by water. Outbreaks of cholera in London in 1831–2, 1848–9, 1853–4, and 1866, each produced considerable loss of life. By the time of the last of these, much of London had been connected to an improved sewage system. The clear spatial correlation of cholera mortality with the parts of the city that had not been connected started to convince the sceptics that there might be a causal connection – proposed by John Snow as early as 1849. However, the prevailing *miasma* theory held that all disease was transmitted through the air that we breathe – a belief to which Florence Nightingale clung right up to her death in 1910.

Quite apart from what we would now immediately identify as a major health issue there was a terrible smell. We must assume that Wordsworth was fortunate in his choice of morning in 1802 to stand on Westminster Bridge admiring the Thames and declaring 'Earth has not anything to show more fair;...The river glideth at his own sweet will'. Sitting in the stuffy Houses of Parliament with the windows closed, a few decades later, was too much. Parliament was always averse to spending money and it was only after the 'Great Stink' of 1858 that a bill was approved giving the necessary authority to create a public health scheme which would transform the city. The engineer most closely associated with the solution is Sir Joseph Bazalgette (1819–91) who, having gained experience in land drainage in Ireland, and railway projects in Great Britain, was appointed engineer to the Metropolitan Sewers Commission in 1852. An article published following an interview with Bazalgette in 1890 suggested that 'If the malignant spirits whom we moderns call cholera, typhus and smallpox, were one day to set out in quest of the man who had been, within the past thirty or forty years, their deadliest foe in all London, they would probably make their way to St Mary's, Wimbledon [Bazalgette's home].'

Bazalgette proposed a series of intercepting sewers on both sides of the Thames at different levels as appropriate to maintain sufficient slope for flow, to fit in with the topography of the city, and to connect with street-based collector sewers. Eventually the intercepting sewers reached locations sufficiently far downstream that the sewage could be pumped into holding reservoirs and discharged into the river at high tide – thus ensuring dispersal out towards the North Sea. The construction of the sewers on the north and south banks of the river was combined with formal reclamation and entrainment to create the Victoria and Albert Embankments, thus at a stroke solving the health problem, removing the smell, cleaning up the river, and creating an improved city environment with riverside parks and roads and underground lines. Bazalgette not only engineered the spatial arrangement of the sewers, but was also careful to specify the materials to be used. Where the gradients were such that the flow rate would be high he specified Staffordshire Blue bricks, baked to exceptionally high temperatures, which would be resistant to erosion. For the mortar to be used to set the bricks lining the sewers he specified the newly developed Portland cement which had been found to be resistant to immersion in water. Because of its somewhat untested use in major public works, he insisted on strength testing of every batch: an early example of strict quality control.

Bazalgette adopted a policy of 'dilute and disperse' for the disposal of the sewage wastes from London. Long sea outfalls were for a time regarded as an acceptable solution to the waste disposal problem – far enough offshore and in deep enough water for any toxicity to be rapidly dissipated. They have somewhat fallen from favour and some form of initial treatment of the waste before the liquids are discharged is now preferred. Treatment implies removal of toxic materials and then adopting a policy of 'concentrate and contain'. Neither is ideal.

Waste disposal

Energy is one of the aspects of infrastructure that we take for granted. Nuclear power is cheap to generate but carries with it the challenge of disposal of the radioactive waste products. It is the continuing heat production of radioactive waste, together with the radioactivity itself, that dictates the engineering requirements of waste disposal facilities. The number one requirement is that there should be no pathway by which radioactivity should be able to escape to reach people – in particular no route by which water supplies might become contaminated. Burial of concentrated waste in the ground seems at first sight to be a rational choice – but what is the effect of temperature rise of the waste material on the enclosing rock or clay? Will cracks form, permitting increased flow of water? There have been major testing programmes in deep stiff clay deposits in Belgium and in extensive crystalline rock formations in Switzerland and Sweden: the study of the effects of heating does not require the use of radioactive waste as a source. The solutions to be adopted are governed more by political decision than by engineering difficulty. Politicians who have to be re-elected every four or five years are not keen to be party to a grown-up decision to store nuclear waste in the area that they represent. The clock ticks and the quantity of waste increases.

The disposal of radioactive waste produces an emotional response because of the insidious and invisible character of radioactivity and the legacy of two bombs dropped in 1945. There are many other substances which are toxic to humans and which form waste products of industrial processes and domestic disposal: for example, heavy metals such as chromium and cadmium are produced in the purification and preparation of metals. They tend not to decay and consequently remain in plant matter to be consumed by birds, fish, or animals and then passed on to man. Domestic electronic devices also contain toxic materials, so the rapid turnover of the consumer society adds to the problems of waste disposal.

The object of controlled disposal is the same as for nuclear waste: to contain it and prevent any of the toxic constituents from finding their way into the food chain or into water supplies. Simply to remove everything that could possibly be contaminated and dump it to landfill seems the easy option, particularly if use can be made of abandoned quarries or other holes in the ground. But the quantities involved make this an unsustainable long-term proposition. Cities become surrounded with artificial hills of uncertain composition which are challenging to develop for industrial or residential purposes because decomposing waste often releases gases which may be combustible (and useful) or poisonous; because waste often contains toxic substances which have to be prevented from finding pathways to man either upwards to the air or sideways towards water supplies; because the properties of waste (whether or not decomposed or decomposing) are not easy to determine and probably not particularly desirable from an engineering point of view; and because developers much prefer greenfield sites to sites of uncertainty and contamination.

Consumer waste represents a luxury that only the developed world thinks it can afford. We dispose of enormous quantities of unnecessary packaging; we are persuaded that the lifetime of computers, televisions, mobile phones, and white goods is short, and the rejected items are discarded. A more serious attempt to separate different types of waste might address questions such as: Can it be burnt? Is it organic and will it decompose? Can it be used as a low grade construction material? Or even, can we reduce the amount of waste that has to be disposed of by more assiduous attention to sustainability and whole-life costing during the design process? Reducing the amount of packaging is one obvious starting point. Legislation requiring manufacturers to take back their products at the end of their useful life would stimulate reuse of materials with associated environmental benefits.

Fortunately, there are signs of increasing consumer concern for recycling and separation and reduction of wastes. The economics of recycling may be uncertain but the heightened awareness of the waste problem has to be beneficial.

Brent Spar

We would not dream of demolishing a Roman antiquity or a Gothic cathedral and dumping it to landfill or turning it into concrete aggregate (I hope), but the planned lifetime of many more recent structures is much less than 100 let alone 1000 or 2000 years. Take offshore oil platforms as an example: once all the oil that can be economically extracted has been removed from the field the structure is redundant. A case history which demonstrates the way in which it is easy to go off the rails with disposal of an offshore structure is provided by the Brent Spar saga.

Brent Spar was jointly owned by Shell and Esso as a floating oil storage container (with a capacity of 50,000 tonnes of crude oil) adjacent to the Brent oil field in the North Sea. It was used from 1976 until 1991 when the completion of a pipeline rendered it redundant. Shell, which had responsibility for decommissioning, conducted an assessment of onshore and offshore disposal options, considered the environmental consequences of these options, and proposed to tow it to a location some 250 km west of the Outer Hebrides, then breach the structure with explosives, and allow it to sink. Various possible outcomes of the explosions were considered – it was known that the steel structure had been somewhat damaged during its lifetime and the effect of the detonations could not be predicted with certainty. For each outcome, Shell estimated the quantities of polluting materials that would be released and the extent and nature of the environmental damage that might be caused. This deep sinking proposal happened also to be the least expensive of the proposals being considered.

However, it is not difficult to turn a considered engineering judgement into a conspiracy – to modify the facts a little and create a publicity stunt – and thus produce a news story which travels rapidly round the world and mobilizes a public opinion which is not too concerned with tracing either the history or the facts of the story. The environmental campaigning organization Greenpeace occupied the platform and claimed that there were 5500 tonnes of oil in the spar – compared with Shell's own estimate of 50 tonnes. (For comparison, *Exxon Valdez* spilled some 40,000–125,000 tonnes of oil into Prince William Sound, Alaska in March 1989.)

Escalating public demonstrations against Shell led to negative publicity and loss of revenue which rapidly became significant relative to the costs of disposal of Brent Spar. In order to regain some moral ground, Shell cancelled the deep sea disposal plans; towed the structure to a Norwegian fjord pending further decisions; arranged for an independent organization, Det norske Veritas (DnV) to review the contents, and organized a public competition to produce ideas for acceptable disposal. DnV confirmed that there were 75–100 tonnes of oil; Greenpeace admitted exaggerating their estimate as part of a more general campaign against any deep sea dumping, and the Spar was cut into rings and used as part of the construction of new harbour facilities in Stavanger, Norway.

There are all sorts of morals to be drawn. The rationality of an engineering argument is not necessarily immediately apparent to the layman and requires careful explanation. The media and the people who read newspapers or watch television like scare stories – these are what are remembered, a correction at a later date will not receive the same publicity. People are very sensitive to suggestions of uncontrolled release of chemicals (or radioactivity, or microorganisms). Transmission of information is today so rapid that what is thought to be a local issue becomes a global issue in a matter of seconds. Engineers need to be prepared to think laterally of the possible unintended consequences of their actions and need to be trained to deal with potentially hostile media. The engineering itself is straightforward.

Chapter 3

'Directing the great sources of power in nature'

> With the civil engineer, more properly so called (if anything can be proper with this awkward coinage), the obligation starts with the beginning. He is always the practical man. The rains, the winds and the waves, the complexity and the fitfulness of nature, are always before him. He has to deal with the unpredictable, with those forces (in Smeaton's phrase) that 'are subject to no calculation'; and still he must predict, still calculate them, at his peril. His work is not yet in being, and he must foresee its influence: how it shall deflect the tide, exaggerate the waves, dam back the rain water, or attract the thunderbolt.

Robert Louis Stevenson often accompanied his grandfather on site visits and appreciated his skills. The Charter of the Institution of Civil Engineers speaks of 'directing the great sources of power in nature'. In late 2010 and 2011 (to take a short sampling period) there were earthquakes in New Zealand and Japan; the Japanese earthquake was followed by a devastating tsunami (tidal wave); there was major flooding in Queensland, Australia, and also in Thailand. Society expects that structures will survive and that the infrastructure will be protected, but can we really 'direct the great sources of power in nature'?

Cnut the Great (king of England AD 1016–1035 and also eventually king of Denmark and Norway) famously (though probably apocryphally) allowed himself to be placed in his throne on the beach of the English Channel near Portsmouth at low tide in order to indicate to his dismayed followers that he – although a king – was unable to prevent the continued advance of the tide. Conditions on that day were probably reasonably calm. It is unlikely that Cnut, realistic about the inexorable nature of the tides, would have expected to be able to exert any control over the weather, but anyone living near the coast is very aware of the awful character of the sea when the tide is high and a storm is brewing. The attempts of man, whether a king or not, to demonstrate some dominance over nature can then seem particularly puny.

There are regularly more or less serious floods in different parts of the world. Some of these are simply the result of unusually high quantities of rainfall which overload the natural river channels, often exacerbated by changes in land use (such as the felling of areas of forest) which encourage more rapid runoff or impose a man-made canalization of the river (by building on flood plains into which the rising river would previously have been able to spill) (Figure 19). Some of the incidents are the result of unusual encroachments by the sea, a consequence of a combination of high tide and adverse wind and weather conditions. The potential for disastrous consequences is of course enhanced when both on-shore and off-shore circumstances combine.

In May 2011, the US Army Corps of Engineers opened sluices or breached levees on the Mississippi in order to reduce flood levels in the river and allow the water to flow into the adjoining floodplains (inundating farm land and small settlements) rather than flooding cities downstream: not an easy decision to make, balancing costs and benefits. Folk memory for natural disasters tends to be quite short. If the interval between events is typically greater than, say, 5–10 years people may assume that such events

19. The 72 m high Clywedog Dam in mid-Wales is formed from a row of 11 concrete buttresses. The level of the reservoir is controlled to smooth the peaks and troughs of flow in the upper River Severn (of which the Afon Clywedog is a major tributary) and reduce the likelihood of flooding

are extraordinary and rare. They may suppose that building on the recently flooded plains will be safe for the foreseeable future.

Flood

'Once did she hold the gorgeous East in fee;...Men are we, and must grieve when even the Shade of that which once was great is pass'd away.' Wordsworth regretted the subjugation of the Venetian republic following the depredations of Napoleon and his armies, but he might have had a similar reaction today if he had been there during one of the periods of high water (*acqua alta*) which submerge extensive regions of the city with some regularity. Construction of a protective barrier to keep high water in the Adriatic from the Venice Lagoon is in progress, but the city will continue its steady stately settlement partly exacerbated by earlier pumping of water from a deep aquifer in order to supply the adjacent industrial area of Marghera, and partly the result of continued geological movements and sea level rise associated with climate change. *Acque alte* have become a fact of life in Venice: tides which reach 1.3 m and cover 40–50 per cent of the area of Venice occur roughly once a year. But much higher tides also occur less frequently. The rarer events, like the floods in the North Sea in 1953 and in New Orleans in 2005, can produce much more damage.

North Sea 1953 floods

On the night of 31 January 1953 a combination of high spring tide and a severe depression produced a tidal surge in the southern parts of the North Sea, along the coastlines of south-east England and the Netherlands, which was locally more than 3 m above mean sea level, with waves up to 6 m high. Around 1600 km of flood protection dykes were overwhelmed in England and some 1000 square kilometres of land beyond inundated; 307 people lost their lives. In the Netherlands 1835 people lost their lives and 1365 square kilometres were flooded including about 9 per cent of Dutch farmland.

In the Netherlands, an ambitious flood defence system – *Deltawerken* (Delta Works) – was designed to protect the estuaries of the rivers Rhine, Meuse, and Scheldt against flood events with an annual probability of occurrence of 0.01–0.025 per cent. The last element of this Delta project to be completed was the storm surge barrier Maeslantkering, in the Nieuwe Waterweg, near Rotterdam in 1998. However, the closure of the east branch of the Scheldt in the 1970s and 1980s caused much more concern and international interest. By this time environmental groups were becoming more vociferous and objected to the permanent closure of the river with the changes to the natural habitat that would be caused in the no longer tidal areas. Which takes priority? Birds and flowers – or people and property? It is not possible to produce a precise balance sheet of costs and benefits: what is the value of a stretch of tidal mudflat? We can probably put a fairly accurate value on the property and agricultural land that might be damaged by flooding – but putting a precise figure on the loss of human life? These become political issues which form elements of environmental impact assessment for engineering projects.

So far as the Oosterschelde (eastern Schelde) was concerned, the most popular environmental choice was to provide a structure with a series of 62 sluice gates each 42 m wide, closed only when adverse weather was expected, spanning between concrete pillars, each placed on specially prepared fabric mattresses filled with sand and gravel on the seabed. The foundations have to be designed to withstand the substantial out-of-balance water pressure when a tidal surge is being resisted, together with repeated wave loading with a frequency of the order of 1 Hz which could *liquefy* the sand. Think back to your experiences on the seashore and the way in which saturated sand, shaken by your foot, rapidly loses its strength and seems to become a liquid: this is another potentially damaging phenomenon which has its origin in the particulate nature of soils.

The tidal habitat has been preserved, but the blockage of even part of the section of the river channel by the piers of the barrier has changed the flow regime to such an extent that movements of sand and shingle by the sea along the coast of the Netherlands have been affected. The movement – and deposition – of offshore sediments is very sensitive to details of the shape and form of the coastline.

In the UK, after the 1953 flooding, protection against recurrence was provided through major investments in raised coastal embankments and through the construction of the Thames Barrier – a miniature version of the Oosterschelde structure – with four 60 m and two 30 m navigable openings which can be closed by rotating, using buoyancy, segmental steel gates which normally sit on the bottom of the river, full of water. The detailed planning of this barrier was delayed because of the need to maintain free access for substantial cargo ships to the main docks of the Port of London. With the expansion in the 1960s of Tilbury Docks downstream of the chosen barrier site, the requirements for the provision of navigable channels were reduced. These investments are intended to protect against tidal surge events with a return period of 1000 years – an annual probability of 0.1 per cent.

Most of the land in the coastal areas of the south-east of the UK is above sea level. These are areas of comparatively recent geological age consisting of deposits of stiff clay (which has been densified by the weight of overlying glaciers and rocks which have now been eroded away), and weak rocks; the chalk which forms the white cliffs of Dover and other landmarks is typical. Erosion of the coastline through the constant battering of the sea and its waves continues. The sea is not easily subdued. The interaction of sea and shore would naturally lead to a coastline quite different from that with which we are presently familiar. A confident civil engineer might attempt to hold the sea back using hard structures, shaped to throw the energy of the water back offshore. Such structures are costly and require regular maintenance in response to the ravages of the sea.

Structures projecting out to sea from the coastline, such as groynes or breakwaters, may be completely impermeable to the passage of water and may provide protection to a harbour from the prevailing currents. Alternatively, they may be deliberately porous: piles of rock or cast concrete units, or even open timber structures. Breakwaters are intended to reduce the energy of the sea and so deposit sediment (sand and shingle). Attempting to tamper with nature by encouraging deposition at one location is to prevent deposition (encourage erosion) somewhere else. The offshore environment is very dynamic and interactive.

Where the density of coastal population is relatively low, 'managed retreat' may provide a more economic (and fatalistic) response to the unequal battle between sea and shore. Isostatic uplift – the natural rebound of the earth's surface as overlying ice melts and/or rock erodes – together with increases in sea level resulting from anthropogenic or naturally occurring changes in global climate – expose to the sea soils and rocks which would naturally be continually removed. Why resist? Salt marshes, flooding land previously reclaimed from the sea, provide rich habitats for wildlife, a luxury of the developed western world. The loss of a few coastal settlements is hardly a new phenomenon. Unnatural man-made hard coastal defences are replaced with soft natural means of absorption of wave energy to improve protection of areas further inland. The coastal sea regains its natural mechanisms of sediment transport and transfer. Evidently, the deliberate sacrifice of coastal property in the interests of long-term economy is socially controversial. In regions of the world – such as Bangladesh – with extremely high densities of population in vulnerable low lying coastal areas the possibilities of preserving existing settlements as sea levels inexorably rise seem bleak.

New Orleans 2005

It wasn't the miles of buildings stripped of their shingles and their windows caved in or the streets awash with floating trash or the live oaks that had been punched through people's roofs. It was the literal powerlessness of the city that was overwhelming. The electric grid had been destroyed and the water pressure had died in every faucet in St Bernard and Orleans parishes. The pumps that should have forced water out of the storm sewers were flooded themselves and totally useless. Gas mains burned under water or sometimes burst flaming from the earth, filling the sky in seconds with hundreds of leaves singed off an ancient tree. The entire city, within one night, had been reduced to the technological level of the Middle Ages.

The disaster that struck New Orleans in 2005 provided the factual background for a gritty novel by James Lee Burke more extraordinary than any fictitious description.

New Orleans is located in the Mississippi River Delta on the banks of the Mississippi River and south of Lake Pontchartrain, 170 km upriver from the Gulf of Mexico. Originally located on the natural levees and high ground along the Mississippi, the city was given permission early in the 20th century to put into effect a drainage plan for the swamp surrounding the city in order to permit expansion and, it was hoped, economic growth. Pumping for drainage increases the stresses in the soil, leading to subsidence of the drained ground, as in Venice. Some parts of the city are as high as 6 m above sea level but the average elevation is about 0.5 m below sea level: some locations are as low as −3 m. Climate change produces a small annual rise in sea level, and there is continuing lowering of the level of the delta sediments as a result of both continued pumping and also the loss of organic surface materials through oxidation by exposure to the air. Ironically, before many of the protective flood control structures were constructed the Mississippi would have kept the delta topped up with new sediments to counter some of the subsidence. Such are the unintended consequences of human intervention.

New Orleans has always been vulnerable to flooding. Flooding caused by Hurricane Betsy in 1965 killed 76 people. The subsequent Flood Control Act of 1965 required the US Army Corps of Engineers to design and construct flood protection in the New Orleans metropolitan area, but in 2005 completion was forecast for 2015. Heavy rain in May 1995 revealed the inadequacy of the pumping system. Both of the elements – protection and pumping – which might have reduced the extent of flooding were in an uncertain state when Hurricane Katrina arrived on 29 August 2005, producing intense winds and a tidal surge.

The failure of most of the protective levees throughout the metropolitan area left 80 per cent of the city underwater – in places to a depth of 4.5 m. Some failed because the embedment of the walls was insufficient to withstand the very high water pressures; others were overtopped for a sufficient time to permit erosion of the banks of the levee. Around 1500 people died.

As with other natural disasters – such as earthquakes – it was not just the damage to property but also the destruction of the local infrastructure of a major city which shocked the world. Failure of power supplies and power lines halted many communications systems; most roads out of the city were damaged; local television and newspapers were disrupted; emergency services and doctors and hospitals were unable to function; police forces were unable to prevent widespread looting. Everything that we regard as part of the normal infrastructure for civilized urban existence was disrupted in a way that revealed the lack of robustness of the systems: a failure in any one location had implications that extended over a disproportionately larger area.

The assessment of the probability of occurrence of extreme events is not easy – by definition there will be few data points. Katrina was an event with a return period of 100–400 years – an annual probability of occurrence of 0.25–1 per cent. Reconstruction and improvement works since Katrina have improved the protection of New Orleans against events with a return period of roughly 100 years and will reduce the impact of rarer more serious events.

The great sources of power in nature are there and will continue to overwhelm man from time to time. We can design to resist these natural phenomena – building on the experience of past magnitudes of earthquakes, typhoons, floods – and constantly updating our statistics of occurrence to improve the accuracy of our forecasts of the likelihood that events of a particular magnitude such as intensity of earthquake, wind speed, rainfall, or flood level, might occur. But an estimate of the probability of occurrence of an event does not define the frequency with which such events will occur. A 200 year flood may occur *on average* once every

200 years, but the probability that it will occur in each successive year is the same – and it could indeed occur each year in succession for many years without upsetting the statistics which relate to the probability of occurrence over hundreds or thousands of years. Risks can be reduced and delayed but never eliminated. It is important that the nature of such risks should be both advertised and understood. The North Sea flooding in 1953 led to the reactive construction of flood protection systems round the coasts of the Netherlands and the south-eastern UK, but the flooding produced by Katrina in New Orleans gives an illustration of the speed with which everything that we take for granted in the infrastructure of our 21st century civilized life can be eliminated.

Earthquake

Major earthquakes occur every year in different parts of the world. The various continents that make up the surface of the Earth are moving slowly relative to each other. The rough boundaries between the tectonic plates try to resist this relative motion but eventually the energy stored in the interface (or geological fault) becomes too big to resist and slip occurs, releasing the energy. The energy travels as a wave through the crust of the Earth, shaking the ground as it passes. The speed at which the wave travels depends on the stiffness and density of the material through which it is passing. Topographic effects may concentrate the energy of the shaking. Mexico City sits on the bed of a former lake, surrounded by hills. Once the energy reaches this bowl-like location it becomes trapped and causes much more damage than would be experienced if the city were sitting on a flat plain without the surrounding mountains.

Designing a building to withstand earthquake shaking is possible, provided we have some idea about the nature and magnitude and geological origin of the loadings. Taiwan is a seismically active country – the magnitude 7.6 Chi-Chi earthquake of 21 September 1999 killed 200 people. The Taipei 101 building (Figure 20) in Taiwan was, for six years from 2004, the tallest building in the

20. The skyscraper known as Taipei 101 contains a *tuned mass damper* in the form of a gold-painted sphere of steel plates hung from steel cables and connected to the building through shock absorbers. Its natural frequency matches that of the building itself and helps to improve structural response under wind or earthquake loading

world. One way to counter the oscillation of a somewhat flexible building induced by ground movement in an earthquake is to include a *tuned mass damper* as a passive means of extracting energy from the movement of the building. The tuned mass damper is a heavy pendulum at the top of the building which is tuned to a frequency close to that of the building and is connected through dampers (shock absorbers). The Taipei 101 damper takes the form of a golden sphere of mass 730 tonnes, built up from steel plates, suspended from eight steel cables, and connected through eight dampers. It is very much on display to visitors who come to the viewing galleries around the 90th storey and who are encouraged to look inwards as well as outwards. It provides an opportunistic but low key example of the importance (and beauty) of civil engineering and the contribution that it makes to the safety of the built infrastructure.

Earthquakes in developing countries tend to attract particular coverage. The extent of the damage caused is high because the enforcement of design codes (if they exist) is poor. The earthquake in Haiti in January 2010 was a magnitude 7 quake with horizontal ground accelerations up to about 45 per cent of earth's gravity. It demolished schools, houses, offices, hospitals, government buildings, parliament, bridges, power stations, and water supply structures. It inevitably had a devastating effect on the life of all people in the country. Estimates of the death toll vary hugely – the number killed was probably between 50,000 and 100,000. The Loma Prieta earthquake in San Francisco in October 1989 was of slightly higher magnitude 7.1; 37 people were killed and there was damage to buildings and infrastructure but at a greatly reduced scale by comparison with Haiti. The geological fault ruptures for the two quakes were broadly similar but there were strongly contrasting effects. California has long had very detailed requirements for the design of all sorts of structures to survive earthquake loading, which are regularly updated after every earthquake occurrence, and these requirements are strictly enforced. The majority of the damage in Haiti was the result of poor construction and the total lack of any building code requirements.

Almost every major earthquake in any part of the world produces some effects that had not been directly anticipated in the development of the building codes for that or other areas. For example, the magnitude 6.8 Kobe earthquake of January 1995 combined the usual horizontal shaking (peak ground accelerations as high as about 90 per cent of earth's gravity) with unusually high levels of vertical shaking (peak vertical accelerations up to about 46 per cent of earth's gravity). The peaks of horizontal and vertical acceleration do not necessarily coincide. But if they do, a bridge which feels only about half its usual weight (because the ground is dropping away) is given a sideways kick. The most striking image from the Kobe earthquake is of the elevated Hanshin highway on its elegant piers which had all toppled sideways. This led to changes in the design codes in Japan and elsewhere to require allowance for such devastating combinations of ground motion.

The earthquake in Haiti is just one of many earthquakes in developing countries where the enforcement of codes is not considered important and the exigencies of daily life are challenge enough without having to worry about building to resist more or less infrequent earthquakes. There is a challenge to developed countries to design building styles which, while only subtly different from traditional styles, nevertheless improve the chances of earthquake survival. Heavy mud or tile roofs on flimsy timber walls are a disaster – the mass of the roof sways from side to side as it picks up energy from the shaking ground and, in collapsing, flattens the occupants. Provision of some diagonal bracing to prevent the structure from deforming when it is shaken can be straightforward. Shops like to have open spaces for ground floor display areas. There are often post-earthquake pictures of buildings which have lost a storey as this unbraced ground floor structure collapsed.

The energy in the shaking ground is large, and the movements and location of earthquakes unpredictable. The very least we can do is to study local geology to spot geological faults which look as though they are storing up energy to be released in a future slip. We can also learn from past events and apply the knowledge gained to the improvement and enforcement of appropriate standards of construction.

Wind

The deck of a suspension bridge has simply to span between the hangers that link it to the suspension cables. However, while the weight of the deck and the traffic that it carries may be the dominant load on the bridge, there may well also be significant horizontal wind loading transverse to the line of the bridge.

The Brooklyn Bridge in New York and the Golden Gate in San Francisco (Figure 21) have decks (and towers) which are formed of

21. The Golden Gate Bridge in San Francisco has steel towers and truss deck which have to resist the lateral buffeting resulting from wind loading

steel lattices. The Tacoma Narrows bridge famously demonstrated the potential for interaction of wind with an intervening bluff body – one that makes no attempt to encourage the smooth flow of the air. Opened in 1940, and known as 'Galloping Gertie', it lasted just four months before it twisted itself to destruction as a result of resonant torsional oscillations generated by wind-induced vortices.

The rules of structural mechanics that govern the design of aircraft structures are no different from those that govern the design of structures that are intended to remain on the ground. In the mid 20th century many aircraft and civil structural engineers would not have recognized any serious intellectual boundary between their activities. The aerodynamic design of an aircraft ensures smooth flow of air over the structure to reduce resistance and provide lift. Bridges in exposed places are not in need of lift but can benefit from reduced resistance to air flow resulting from the use of continuous hollow sections (box girders) rather than trusses to form the deck. The stresses can also flow more smoothly within the box, and the steel be used more efficiently. Testing of potential box girder shapes in wind tunnels helps to check the influence of the presence of the ground or water not far below the deck on the character of the wind flow.

Built within a very few years of each other, the suspension bridges across the Forth (near Edinburgh, 1006 m span, completed in 1964) and the Severn (near Bristol, 990 m span, completed in 1966) adopted the contrasting deck forms of truss (Forth) and aerodynamically shaped box (Severn). The rapid adoption of box girders turned out to be a jump in technology which in fact required extensive further development. Collapses of box girder bridges in Wales and Australia while they were being constructed in the early 1970s indicated a need to understand more carefully how the stresses would flow round the irregular cross-section of the box. Hollow sections (even aircraft fuselages) require some internal stiffening (shear stiffening) to discourage them from changing in shape under load (just as the storeys of a building require shear stiffening to prevent them collapsing – changing shape – in an earthquake). A truss gives a clear picture about the route that the internal forces must take. Box girder bridges are under particular risk during construction. The failures that occurred prompted rapid research and introduction of design guidance. This then led to a major programme of strengthening (increasing the weight) of most of the bridges of this type that had been built in the UK in order that they could be deemed safe under the most critical (and increasing) traffic and environmental loads.

Harnessing the sources of power in nature

The damage that can be inflicted by natural movements of wind, water, and earth is enormous. We can design structures to resist or survive these movements, but do we choose to make the structure heavy and stiff so that it cannot move, but generates high internal stresses, or rather to make the structure flexible so that it goes with the flow, generating rather low internal stresses but experiencing larger movements which may be unpopular with the users of the structure?

We can also attempt to exploit these natural sources of power to contribute towards the ever increasing demands of society for energy. Almost every method of energy generation has its disadvantages. Coal is seen as a dirty fuel contributing to the build-up of carbon dioxide in the atmosphere with attendant effects on climate change. There are ways of reducing the output of carbon dioxide at the power station but these obviously put up the cost. Oil and gas are seen as cleaner fuels but their scarcity and location often in areas of the world which are not noted for their political stability suggest that long-term reliance on them as fuels is risky. Nuclear power was hailed after the Second World War as the answer to the demands for cheap energy, but the politics of disposal of spent nuclear fuel have not been resolved. Even if engineering solutions can be found, the public remains sceptical.

Hydroelectric power is a clean way of generating electricity from the reliable combination of rainfall and gravity. But today there is great concern to prevent the damage to the environment that is caused by the combination of flooded valley and controlled, possibly reduced, river flow downstream of the power station and the removal of a sometime waterfall into pipework and turbines.

Wind has for many centuries been seen as a desirable natural source of power to be tamed for the use of mankind. There has recently been a massive development of onshore and offshore wind farms in many parts of the world – groups of tens or hundreds of wind turbines generating considerable quantities of power (when the wind blows). Opinions differ as to their aesthetic contribution: the windmills (or more often wind pumps) built in northern Europe in the 17th and 18th centuries that are now seen as rather romantic relics in a rural landscape may then have been regarded as visual intrusions, notwithstanding their contribution to pumping of water, drainage of land, grinding of corn. There is a certain elegance in a group of slender modern wind turbines rotating gently – but perhaps as a distant view and not on one's doorstep. Offshore locations may be seen as further from anyone's backyard: the wind is stronger but there are other environmental loadings from the surrounding sea as well.

Light shining on photovoltaic cells generates electricity from silicon semiconductors – there are solar photovoltaic installations at all scales from providing illumination for individual traffic signs to powering remote lighthouses, domestic supplies and office buildings. The wind may or may not blow and the sun may or may not shine but the tide will come in and out twice a day without fail. Electricity can be generated from the tide by installing a dam across an estuary and allowing the water to flow through turbines from one side to the other depending on the state of the tide. The Rance tidal power station in northern France was the world's first, opened in 1966. The tidal range is about 8 m and the power generating capacity is 240 MW with a 700 m dam across the estuary. As with on-shore hydroelectric power schemes there are environmental concerns with loss of habitat for birds and plants in an estuary where there would have been extensive mudflats.

The sea itself has a lot of energy in addition to the effects of the tides. Tidal current devices can be completely invisible beneath the surface of the sea – the tidal currents are encouraged to flow through turbines tethered to the seabed. Wave-power has been studied for several decades; various likely candidates have emerged but the extent of actual deployment has been modest. The invention of the devices themselves is primarily a mechanical challenge. The civil engineering challenge to anchor them to the seabed is significant: the power of the sea to generate electricity is also the power that batters the coast in stormy weather and destroys puny man-made structures that are placed in its path. The anchorage has to resist the very natural power that the devices are trying to harness. The need for interdisciplinary awareness is evident.

Chapter 4
Concept – technology – realization

Engineering is concerned with finding solutions to problems. The initial problems faced by the engineer relate to the identification of the set of functional criteria which truly govern the design and which will be generated by the *client* or the promoter of the project. A client who says 'I need a road bridge across this fjord' may really be saying 'I need some form of transport connection between the communities on each side of the fjord' or 'I need the communities on both sides of the fjord to have access to facilities which are presently only available on one side'. The first description of the problem seems to exclude the possibility of a tunnel or that a rail link might be more appropriate. The second description assumes that there is a need for people and vehicles to travel to and fro. The third description gives much more freedom to explore what the real needs might be: a *solution-neutral problem definition*. The more forcefully the criteria are stated the less freedom the design engineer will have in the search for an appropriate solution.

Design is the translation of ideas into achievement. Figure 22 shows the broad framework of the elements of the design process. The *designer* starts with (or has access to) a mental store of solutions previously adopted for related problems and then seeks to compromise as necessary in order to find the optimum solution satisfying multiple criteria. The design process will often involve iteration of *concept* and *technology* and the investigation of

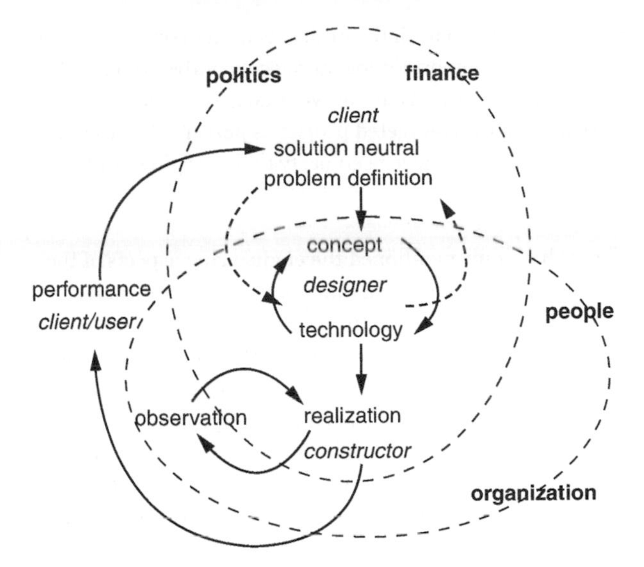

22. Elements of civil engineering design process

radically different solutions and may also require consultation with the *client* concerning the possibility of modification of some of the imposed functional criteria if the problem has been too tightly defined.

The term *technology* is being used here to represent that knowledge and those techniques which will be necessary in order to realize the *concept*; recognizing that a *concept* which has no appreciation of the *technologies* available for construction may require the development of new *technologies* in order that it may be realized.

Civil engineering design continues through the *realization* of the project by the *constructor* or contractor. A loop of *observation* has been attached to the stage of *realization* to indicate that there may well be assumptions made at the design stage – for example, details concerning the strata and properties of the ground – which cannot be confirmed until the project is under construction but which are rather important for the success of the project. The process of design extends to the eventual assessment of the *performance* of the completed project as perceived by the *client* or *user* (who may not have been party to the original problem definition).

So far we have only mentioned the engineering aspects of the process. However, there are a number of non-technical factors which will also play a role in the success of the project: the *political* and *financial* environment must be supportive; the *organizational structure* and the skills of the *members of the team* of designers and constructors need to be appropriate. The dotted ellipses in the diagram are purely schematic – organizational factors will affect the interaction of client and designer and constructor, but will probably be most important within the construction team itself.

We will look at a number of projects against this framework and try to draw out elements of *concept*, *technology*, *organization*, *politics*, and *finance* which have influenced their eventual outcome. It is rare for all these elements to be satisfactorily interwoven. The interplay of *concept*, *technology*, and *realization* to achieve an optimum outcome is often hidden behind the resulting structure. In some of the projects the *political* will for completion has been more or less sustained (with moments of faltering). The absence of continuous *political* support is not necessarily fatal, but it exposes the project to weaknesses or lack of *robustness* in the elements of *organization* and *people* – and that is when failure can occur.

Sydney Opera House

There is a link between the construction of the Sydney Opera House (Figure 23) and the construction of the dome of Santa Maria del Fiore in Florence by Filippo Brunelleschi. In 1420,

23. The roofs of the Sydney Opera House as we see them today are the result of iteration between architect and civil engineer to find a way in which the original competition-winning sketch could be turned into a structurally tractable reality

Brunelleschi committed himself to the realization of a design *concept* produced in 1367 by Neri di Fioravanti for the Opera del Duomo, a group largely composed of members of the wealthy Wool Merchants' Guild who knew about wool but had no idea how the dome shown in the model produced by Neri should be built; modification of the appearance was not an option. The *technology* had to be found to realize a *concept* which had been frozen many years before. At Sydney too the engineering challenge was to produce a structure designed by someone who had strong views on appearance and form but no detailed knowledge of how that appearance and form could in fact be achieved: the link between *concept* and *technology* did not exist.

Sydney Opera House is one of the most memorable buildings of the 20th century. It provides an icon for Sydney and its location on Bennelong Point, projecting out into Sydney Harbour, ensures its visibility and its central presence in pictures of the city. The appearance of the building represents the realization of the vision of the Danish architect Jørn Utzon, whose entry in the competition organized by the government of New South Wales was awarded the first prize in 1957. The specification for the competition was provided by a committee who had general ideas of the needs for venues for opera, concert, and theatre performances in the city but who had no experience of managing the construction of a building of this complexity. Utzon's schematic proposals did not actually meet all the requirements of the competition but showed an imagination in their external appearance which was largely absent from the majority of other entries. Utzon's striking sail-like roofs exploited the prominence of the site, but bore no relation to the functions of the performance spaces that they contained.

Ove Arup was a 20th century structural engineer who recognized the supreme importance of working on an equal footing with architects and other building specialists (such as mechanical and electrical engineers responsible for the building services – heating and ventilation, security and control systems, communications, etc.). The engineering company that carries his name is known across much of the globe. If informed members of the public were asked to name just one civil/structural engineer of the 20th century would Arup be the most likely name?

Ove Arup was born to Danish/Norwegian parents in Newcastle in 1895 and received his early schooling in Hamburg. The widespread chalk geology of Denmark was ideal for manufacture of cement. The Danish company Christiani & Nielsen, established in 1904, specialized in the design of reinforced concrete for underwater structures. In 1922, they recruited Ove Arup as a graduating engineer from the Copenhagen Polytechnic. Sent to work in Hamburg he sought out architects such as Le Corbusier – who proposed that true architecture was found only in the work of engineers because they followed the laws of nature and pursued the goal of utility – and Gropius – whose Bauhaus School in Weimar was founded in 1919 with the vision of bringing together craftsmen, artists, architects, and technology to pursue a utopian vision for the world in the aftermath of the Great War.

Arup believed that the best architecture in reinforced concrete is to be found among large engineering structures. When he created his own consulting company after the Second World War he sought out architects with whom to collaborate and subsequently set up his own architectural practice within the Arup group. 'The paramount importance of getting the right design is hardly understood by laymen (including clients), rarely grasped by building authorities, often not by architects, who leave the costing of jobs to quantity surveyors, or even by consulting engineers who concern themselves with structural stability but leave matters of construction to the contractor.'

Leslie Martin and Eero Saarinen, international members of the competition panel, recommended that Utzon would need technical assistance, and Arup's firm was duly appointed by the government of New South Wales to oversee the engineering aspects of the project. Over the first few years the link between Danish architect and Danish engineer was generally fruitful.

An arch or vault curved only in one direction transmits loads by means of forces developed within the thickness of the structure which then push outwards at the boundaries. A shell structure is a generalization of such a vault which is curved in more than one direction. An intact eggshell is very stiff under any loading applied orthogonally (at right angles) to the shell. If the eggshell is broken it becomes very flexible and to stiffen it again restraint is required along the free edge to replace the missing shell. The techniques of prestressing concrete permit the creation of very exciting and daring shell structures with extraordinarily small thickness but the curvatures of the shells and the shapes of the edges dictate the support requirements.

To sketch a shape is not to be able to develop a structure. At the time of the Sydney Opera House competition, computer abilities were limited, and the nature of double curvature surfaces that could be analysed was usually confined to simple shapes. Utzon had not defined the geometric shape of his roofs and had no prior experience of building such complex structures. His initial aspiration had been to cover the various buildings of the Opera House complex (two theatres and a restaurant) with thin concrete shell roofs. However, when analyses of shapes similar to those sketched in the competition entry were performed it became clear that the stresses in the shells envisaged and the loads required to support these shells along their boundaries and at the points of support would be excessive. The freely varying curvatures produced no serious possibilities of standardization. The solution finally adopted iteratively through partnership between Utzon and Arup was to replace the thin shells by series of concrete ribs which could be analysed as individual arches, and to form the external shape of all the roofs from segments of a sphere of a single radius. This immediately eased the description of the geometry and opened up possibilities of an effective production and erection procedure: an eventual harmonization of *concept* and *technology* for the design of the roofs.

Phase I, the construction of the platforms which provide the access to the theatres and house the below-stage facilities and the hollows into which the auditoria would be fitted, was begun in 1959 before the details of the roofs or of the auditoria had been chosen. There was a *political* reason for this: an impending change of government looked quite likely to result in abandonment of the project. Construction of an Opera House in Sydney was not universally popular and uncontrolled overruns in both time and budget were predicted. Naturally subsequent modifications to the phase I work were made to accommodate the changing ideas introduced in phase II – the roofs – and the eventual phase III – the construction of the performance spaces themselves.

The roofs were constructed over the period from 1962 to 1966. The external appearance of the opera house is the consequence of covering the roofs with precast panels, each comprising a large number of bright white ceramic tiles. The need for each of these panels to be dimensioned and manufactured to fit precisely in its final destination produced further challenges of construction and engineering.

Utzon was dismissed from the project in 1965–6 and the third phase, which was concerned with the internal arrangements of the several structures and the arrangement of seating in the auditoria and the closure of the spaces created by the roofs, was handled by a substitute team of Australian architects. The Opera House was eventually opened by Queen Elizabeth in 1973: the final cost was some A$100 m compared with an initial estimate of A$2.5 m in 1957. There are those who believe that Sydney was denied the pure Utzon masterpiece that could have been. The iconic Opera House that we see externally today represents a product of a somewhat turbulent partnership between an architect, pursuing an aesthetic ideal *concept*, and an engineer concerned always to find the *technology* to be able to turn that *concept* into a viable structure.

The adequacy of the necessary non-technical elements fluctuated. *Political* support came and went as governments changed, but once a major start had been made it was difficult to turn back, even if the party in power at any time was reluctant to find yet more *finance* to keep the project going. Utzon was driven off the job to some extent by the delays in paying for what he saw as entirely appropriate design activities. The *organizational* structures were not satisfactory and the breakdown in confidence among the various *people* involved was partly the result of confusion over the responsibilities assigned to Utzon and to Arup. We conclude that none of the four non-technical elements was really satisfactorily in place.

Bell Rock lighthouse

The sea has been available for ever as a medium of transport. In the 19th century it was quicker to travel from Rome to Ancona by sea round the southern tip of the boot of Italy (a distance of at least 2000 km) than to travel overland, a distance of some 200 km as the crow flies. Land-based means of transport require infrastructure that must be planned and constructed and then maintained. Even today water transport is used on a large scale for bulky or heavy items for which speed is not necessary.

However, the sea is not without its hazards and, even in clear weather, rocks or skerries provide an obvious danger to shipping especially if they cannot be seen above the surface of the sea. The Inchcape Rock some 18 km off the east coast of Scotland was long recognized as a notable hazard to shipping aiming for the ports of Dundee or Edinburgh. In the 14th century, the public-spirited Abbot of Arbroath (Aberbrothok – where the declaration of Scottish independence was signed in 1320) was sufficiently concerned at the wreckage that he had a bell fixed to the rock which would toll in the waves. In Southey's ballad the wrecker Ralph the Rover is himself wrecked on the rock from which he had previously gleefully removed the bell.

By 1800, the need to regularize the marking of navigation hazards was becoming clearer. In Scotland, the Northern Lighthouse Board, and, in England and Wales, Trinity House were charged by Act of Parliament with the construction of lighthouses around the coast. Of the Scottish lighthouses the majority were designed, and their construction managed, by members of the Stevenson family which maintained this specialist reputation across many generations. When Japan opened up to the west after the Meiji revolution of 1868, Richard Henry Brunton, sent out by the Stevensons, became known as the father of Japanese lighthouses, and modelled their organization on the Northern Lighthouse Board.

The Stevenson family is better known today for the writer Robert Louis Stevenson who actually started an education in civil engineering at Edinburgh University. He always had immense respect for the profession of his father and grandfather:

Say not of me that weakly I declined
The labours of my sires, and fled the sea,
The towers we founded and the lamps we lit,
But rather say: In the afternoon of time
A strenuous family dusted from its hands
The sand of granite, and beholding far
Along the sounding coast its pyramids
And tall memorials catch the dying sun,
Smiled well content, and to this childish task
Around the fire addressed its evening hours.

Our knowledge of the details of the construction of the Bell Rock lighthouse (Figure 24) comes most eloquently from the edited entries of the diaries of his grandfather, Robert Stevenson. Robert Stevenson (1772–1850) was apprenticed to his stepfather, Thomas Smith, and gained experience in the construction of several lighthouses around the coast of Scotland from 1796 to 1802. He succeeded Smith as Engineer to the Northern Lighthouse Board in 1808, having married his stepsister and become Smith's business partner in 1802. Though the Bell Rock lighthouse was not the first lighthouse on which Robert Stevenson had worked, its construction was certainly the most challenging and its successful completion established his position as a leading lighthouse engineer. Here we come across the uncertainties of the historical record.

Who designed the Bell Rock lighthouse? John Rennie, who was some 10 years older than Robert Stevenson, was appointed chief Engineer for the project by the Northern Lighthouse Board. Both Rennie and Stevenson had submitted designs for the lighthouse which were certainly inspired by the Eddystone lighthouse designed by John Smeaton and completed in 1759.

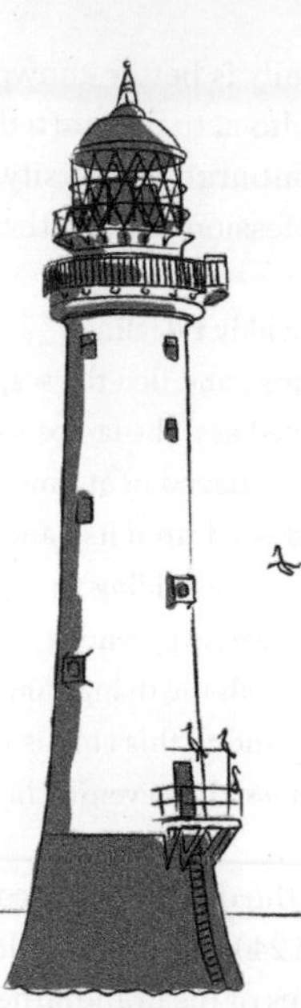

24. The successful construction of the Bell Rock lighthouse on a rocky skerry which is submerged except at low tide was a triumph for both John Rennie and Robert Stevenson

Stevenson was the 'resident engineer' of the project: he took daily decisions, shared many of the privations of his labourers on the rock and, in a sense, 'built' the lighthouse. Subsequent accounts deliberately leave either Rennie or Stevenson out of the picture. In fact, Stevenson was regularly seeking Rennie's approval and Rennie made design recommendations concerning the profile of the tower and the importance of dovetailing the masonry. The *concept* and much of the *technology* must be attributed to Rennie (with inspiration from Smeaton). The *realization* was more in Stevenson's hands. As Chief Engineer, Rennie would have had to take responsibility for any failure of the project. He was very aware that Robert Stevenson was vigorously trying to establish a business and reputation as a designer of lighthouses: he needed to emphasize his own contribution. Rennie had plenty of other demands on his design experience. The design and construction of the Bell Rock lighthouse was the product of this (sometimes uneasy) partnership.

The Inchcape Rock is only exposed at low tide – at high spring tides it is covered by about 5 m of water. Even when the rock is exposed, the rough sea may not permit a landing. An initial period of interrupted working was required merely to establish accommodation on the rock above high water level, thus permitting the workers to remain on the rock for extended periods. The lighthouse was created as a three-dimensional puzzle, with the interlocked stones carefully cut and assembled on land before being shipped out to the site where the rock itself had been carefully prepared to receive the first stones. All the masonry had to fit tightly into the rock and into itself so that it would resist the power of the sea. The access door to the lighthouse was located sufficiently high up on the leeward side of the tower that it would be rarely exposed to the waves and only above this level did the tower contain rooms and connecting stairways. Parliamentary approval for the construction of the lighthouse was given in 1806 – the building was completed in 1811.

Panama Canal

Keats compares the thrill experienced by 'stout Cortez when with eagle eyes he star'd at the Pacific – and all his men look'd at each other with a wild surmise – silent, upon a peak in Darien' with the excitement of discovering Homer for the first time. In fact, it was Vasco Nuñez de Balboa who, in 1513, climbed a peak in the Panama isthmus of Latin America, near the town Santa Maria de la Antigua del Darien that he had recently founded, and saw the Pacific Ocean. He immediately claimed it and its entire coastline for Spain. The realization of the narrowness of the land barrier between the Atlantic and the Pacific obviously opened up the possibility of a waterway to provide a transport passage.

However, the survey report prepared for Charles V (Holy Roman Emperor and King of Spain) soon after Balboa's discovery concluded that such a link could not be built. The *concept* was there but not, at that time, the *technology*.

The Industrial Revolution in Europe spawned a huge flurry of canal building. Canals could either keep to a contour and wind around the hillsides or aim for more direct routes with locks to change level and tunnels, and negotiate geographical obstacles. (It was recognition of the continuity and reappearance of geological strata in the exposed rocks of canal cuttings that led William Smith to develop the first accurate stratigraphical map of the geology of Britain, in 1815.) Some three and a half centuries after Balboa, Ferdinand de Lesseps, fresh from the completion of the Suez Canal in 1869, formed a company, La Société Civile Internationale du Canal Interocéanique de Darien, for the specific purpose of constructing a canal to link the Atlantic and Pacific Oceans through the American isthmus. De Lesseps argued stubbornly that the best option would be a sea level canal. The Suez canal is some 160 km long and at sea level (and cost more than twice the original estimate). De Lesseps had no engineering knowledge; his forte was as an organizer of finance. The isthmus connecting South America and North America is particularly narrow in the region of Panama (at that time part of Colombia), so it was fairly evident that a link between the Atlantic and Pacific Oceans should be built here, giving a saving of some 13,000 km in the journey from New York to San Francisco. But the topography, geology and climate of the Panama region were quite different from the conditions at Suez.

A treaty negotiated with the Colombian government in 1878 granted exclusive rights to the Société to build the canal at Panama. Several schemes devised by the surveyor, Lucien Wyse, were all rejected by De Lesseps because they involved locks and tunnels. Following a second exploration Wyse suggested a sea level canal with a 7.7 km tunnel through the continental divide and it was this canal that was anticipated in the treaty. At meetings in Paris to raise finance, Baron de Lépinay was one of the few present to have experience of engineering in the tropics. His certainty that the sea-level route would involve excessive excavation (and his proposal of a route close to that eventually constructed) went unheeded. De Lesseps was convinced that the *technology* required for his chosen canal scheme would somehow emerge. After several years of work it was recognized after all that a canal with locks was needed. Losing faith in De Lesseps, the shareholders declined to provide further financial support; work ended in May 1889. All assets of the Company were subsequently sold to the United States of America for $40 m.

The USA supported the movement for independence of Panama from Colombia in 1903 and then imposed a treaty on the new Panamanian government transferring the 16 km wide Canal Zone to the USA and the project itself to the US government. The Panama Canal Zone was transferred to Panama in 2000.

Construction began in 1904. An initial effort to control disease in the Canal Zone was seen as essential to preserve the construction crews. The route of the canal followed that of the transisthmian railway which was put to good use in delivering plant and materials and removing spoil. John Stevens (Chief Engineer, 1905–7) championed the design of a canal with three pairs of locks on each side of the continental divide. The summit was crossed through a deep cut, the Culebra Cut, renamed the Gaillard Cut after an engineer who died during construction, and the Gatun Lake formed by dams which when built were the largest in the world (and retained the largest lake). This lake controlled flooding of the rivers in the area, kept the canal full of water with no need for pumping for operation of the locks, and generated electricity sufficient to run the canal and the towing engines. The formation of the Cut, some 12.6 km long, was regularly plagued by landslides: slopes up to 150 m high slid into the canal during construction. Landslides remain a hazard: in 1986 a slope some 115 m high slid into the canal, but it was closed for only 12 hours. Smaller landslides occur rather regularly.

The canal was completed and opened in 1914 with an eventual cost of some $375 m, a little lower than the 1907 estimate, and in spite of the landslides and decisions to deepen the canal. The construction of the Great Pyramid at Giza had a maximum labour force of some 25,000 – the labour force for the Panama Canal was around 40,000 at its maximum with similar organizational challenges of housing, feeding, recreation, and health for the workers and their families.

The Panama Canal is of such economic and strategic importance in its elimination of the need to confront the stormy waters of Cape Horn that naval architects (engineers) have carefully designed 'Panamax' boats which will just precisely fit through the locks of the canal having a beam no greater than 32.3 m and draught 12 m, and pass under the Bridge of the Americas constructed across the Pacific entrance to the Canal in 1962. Bigger boats would require a bigger canal. An attempt to widen the canal in 1940 was abandoned because of international distractions but a current project to expand the capacity of the Canal by providing a third shipping lane with larger locks, with maximum beam 49 m and draught 15 m, and with increased dredging of the navigation channels, is intended for completion in 2014. The maximum cargo capacity of ships which are able to pass through the Canal will more than double.

The original problem definition for the Panama Canal was clear: to provide a link for shipping between the Atlantic and Pacific Oceans. However, De Lesseps defined the problem too tightly, adding the requirement that it should be a sea-level canal for which the *technology* was not available. By the time the mismatch of *concept* and *technology* had been recognized the essential elements of *realization* had unravelled. *Political* support was wavering and the *financiers* had grown impatient through lack of action and results. Once the Americans took over there was a single-minded *political* will to drive the project through, the *political* will drove the provision of *finance*. The *organizational* skills and structures required to complete the project were carried out by civil engineers within the US Army Corps of Engineers.

High speed railways

Brunel's design in the 1830s for the Great Western Railway connecting London with Bristol can be seen as an early example of a high speed railway line. His choice of a broad gauge of 2.1 m at a time when all other railways were being constructed with a gauge of 1.435 m was not very clever – logical (greater comfort and more spacious rolling stock) but too late. In other respects his proposals were visionary – easy gradients and straight alignments, as far as possible.

Since the mid 20th century there has been a realization that high speed rail links can provide faster journey times from city centre to city centre than air travel and with less environmental damage in terms of carbon footprint (carbon dioxide emission) per passenger mile. Typically, high speed rail produces 10 per cent of the carbon emission of air travel for journeys such as London to Paris (340 km) or Edinburgh (530 km), or Tokyo to Osaka (400 km) and 16 per cent of the carbon emission of car travel. High speed rail works well (economically) in areas such as Europe and Japan where there is adequate infrastructure in the destination cities for access to and from the railway stations. In parts of the world – such as much of the USA – where the distances are much greater, population densities lower, railway networks much less developed, and local transport in cities much less coordinated (and the motor car has dominated for far longer) the economic case for high speed rail is harder to make.

The most successful schemes for high speed rail have involved construction of new routes with dedicated track for the high speed trains with straighter alignments, smoother curves, and gentler gradients than conventional railways – and consequent reduced scope for delays resulting from mixing of high speed and low speed trains on the same track (which was the regular problem on the line from the Channel Tunnel into London until the dedicated high speed line HS1 was completed). An attempt in the late 1970s to introduce a high speed Advanced Passenger Train which would run on existing track in the UK, using a clever tilting mechanism, failed after trials with full-scale rolling stock revealed that the existing track separation might lead two tilting trains to touch if the tilting mechanism failed on one of them, and identified issues concerning the electrical pick-up from overhead lines. These essentially mechanical difficulties have been solved, and tilting trains are now widely used to enable higher train speeds on old track. However, the advantage of using specially engineered new lines is clear.

Japan led the way in developing a network of dedicated high speed rail routes for their *Shinkansen* ('new main line' – see Figure 25): a success story subsequently followed in the later 1960s with the development of the TGV (Trains à Grand Vitesse) network in France initially in the form of radial routes from Paris and subsequent links to adjacent countries. Since then high speed rail routes have been constructed in many other parts of the world.

'Is then no nook of English ground secure from rash assault?... ye torrents, with your strong and constant voice, protest against the wrong.' Wordsworth was less than happy about the planned encroachment of the railway network into his beloved Lake District – one of the rural lines that survived the massive cuts of the 1950s and 1960s. Development of such a network is obviously speeded if the constraints on choice of route are reduced. Political requirements for extensive public consultation (where the presumption seems often to be against any development at all) and sensitivity to the effect of the new lines on the environment will hold up design and construction. Straighter routes in hilly terrain inevitably imply tunnels through the hills and viaducts across valleys and these are unavoidably visible. Wordsworth would not have approved.

25. The Japanese high speed train, the Shinkansen, revolutionized rail transport, through post-war infrastructure investment, and provided an example for similar projects in other countries

The Shinkansen project grew out of the reconstruction of Japan after the Second World War. Construction of the first line, the Tokaido linking Tokyo with Kyoto and Osaka (515 km) was begun in 1959 and opened in 1964 at just the time when people were forecasting the demise of long distance rail transport and its replacement by air travel. The new 210 km/h trains were an immediate success, opening the possibility of day trips between Osaka and Tokyo. Other lines followed and the Tokaido was itself extended to Hiroshima and beyond.

Train technology has also evolved and the Shinkansen tracks now typically support three speeds of train: *Kodama* (echo) is the slowest, *Hikari* (light) is next, and *Nozomi* (hope) is the fastest with speeds up to 300 km/h – although much higher speeds have been reached in speed trials on conventional rail systems and in special tests using magnetic levitation to lift the train and eliminate resistance between the train and the rails. The high speed (581 km/h) in these latter trials comes with an even stronger requirement of straight alignment.

A railway is a system of many contributory parts, just as in Brunel's day, but the complexity has increased. The civil engineering content is substantial - trackbed construction and track support - using transverse sleepers or a continuous concrete slab which is more expensive but can permit a shallower track profile which is helpful in reducing the size of tunnels. The track layout may involve substantial construction work in the form of viaducts and tunnels. But the trackbed is only there to support the track (which includes points and crossovers) which is subject to the dynamic loading of the train wheels. And the track is only there to support and guide the trains themselves. Successive trains have to be separated safely so that a signalling or train control system is required: the Shinkansen lines use a centralized computer based control system as more reliable than the familiar trackside signals. Civil engineering, mechanical engineering, electrical power engineering, computer systems engineering, ventilation engineering (tunnels), aerodynamic engineering, acoustic engineering (noise) are all there. These separate areas may be the responsibility of separate design teams but no team can work in isolation and each must at least understand the language and constraints of the other teams. Someone has to ensure that all the elements are indeed compatible to ensure the robustness of the process of design and realization.

The *concept* of the Shinkansen in Japan was of *political* origin after the Second World War as part of the provision of infrastructure to draw together the shattered country. The *technology* of advanced locomotives gradually developed - and is still developing - but the key decision in Japan and France and elsewhere was to build new track. There was a unity of purpose which drove the projects forward, surmounting any local planning difficulties.

Millennium Bridge, London

The Millennium Bridge in London (Figure 26) allows pedestrians to cross the Thames between St Paul's Cathedral and the former South Bank power station which is now the Tate Modern art gallery. This bridge achieved fame on opening because the footfalls of the people crossing the bridge excited a lateral/torsional displacement mechanism which was deemed to be unacceptable, though not actually dangerous. The architect to whom the design of the bridge had been attributed by the press, Norman Foster, rapidly dissociated himself from the engineering design which at least had the benefit of bringing the engineer - Arup - into the public eye (any publicity is good publicity).

26. The Millennium Footbridge across the Thames in London is a suspension bridge with an extremely low ratio of rise to span. In the background, the landmark dome of St Paul's Cathedral conceals clever internal structural details devised by the engineer/architect Christopher Wren

The Millennium Bridge is a suspension bridge with a very low sag-to-span ratio which lends itself very readily to sideways oscillation. There are plenty of rather bouncy suspension footbridges around the world but the modes of vibration are predominantly those in the plane of the bridge, involving vertical movements. Modes which involve lateral movement and twisting of the deck are always there but being *out-of-plane* may be overlooked. The more flexible the bridge in any mode of deformation, the more movement there is when people walk across. There is a tendency for people to vary their pace to match the movements of the bridge. Such an involuntary feedback mechanism is guaranteed to lead to resonance of the structure and continued build-up of movements. There will usually be some structural limitation on the magnitude of the oscillations – as the geometry of the bridge changes so the natural frequency will change subtly – but it can still be a bit alarming for the user. The excitation of lateral modes at a frequency between 0.5 and 1 Hz by synchronous footfall – 'synchronous lateral excitation' – became a problem once there were more than a certain number of people on the bridge.

The Millennium Bridge was stabilized (*retrofitted*) by the addition of restraining members and additional damping mechanisms to prevent growth of oscillation and to move the natural frequency of this mode of vibration away from the likely frequencies of human footfall. The revised design – taking account of specially commissioned research – ensured that dynamic response would be acceptable for crowd loading up to two people per square metre. At this density walking becomes difficult so it is seen as a conservative criterion.

The *concept* for the bridge was clear in the mind of the designer but the particular mechanism that was excited by the crowds on the opening day had been overlooked – the understanding of the *technology* was incomplete. The designing engineer did not know what it was that he did not know. A failure perhaps, but once the bridge had been stiffened and reopened the memory of the original difficulties more or less evaporated.

Chapter 5
Robustness

When we travel by air we take for granted that the sophisticated piece of equipment that is the modern aeroplane will probably be controlled by computer, and naturally suppose that there is a degree of redundancy or robustness in the system so that if the computer fails then there is another one ready to take over the control. The humans in the cockpit provide a last resort. The development of appropriately safe systems requires that the parallel control systems should be truly independent so that they are not likely to fail simultaneously. Robustness is thus about ensuring that safety can be maintained even when some elements of the system cease to operate. Such robustness is obviously extremely important in air travel, railways, nuclear installations, or chemical plants, where the consequences of failure are high in terms of human life or economic consequence but civil engineering systems are also frequently complex. Complex systems can produce unexpected modes of behaviour especially when novel technologies are being adopted. There is a human element in all systems, providing some overall control and an ability to react in critical circumstances. The human intervention is particularly important where all electronic or computer control systems are eliminated and the clock is ticking inexorably towards disaster.

Although ultimately whenever a structural failure occurs there is some purely mechanical explanation – some element of the structure was overloaded because some mode of response had been overlooked – there is often a significant human factor which must be considered. We may think that we fully understand the mechanical operation, but may neglect to ensure that the human elements are properly controlled. A requirement for robustness implies both that the damage consequent on the removal of a single element of the structure or system should not be disproportionate (mechanical or structural robustness) but also that the project should not be jeopardized by human failure (organizational robustness).

Ronan Point: progressive failure and redundancy of design

When Miss Hodge lit her gas cooker at 5:45am on 16 May 1968 in order to make a cup of tea, an explosion occurred because of a gas leak and she was disappointed to find that part of her flat on the 18th floor together with the whole of the rest of the corner of her 22 storey block of flats at Ronan Point in south London collapsed to the ground (see Figure 27). Ronan Point was one of many blocks of flats built as economically as possible to provide cheap housing for people moving to the big city or displaced by slum clearance schemes or as a replacement for housing destroyed in the Second World War bombing. Ronan Point had been constructed in 1966–8 using what seemed to be a very efficient form of construction in which individual precast concrete panels were brought in and lifted into place using cranes and then connected together in a modest sort of way along their edges. It was rather like a house of cards and, like a house of cards, when the explosion occurred (Miss Hodge had failed to notice the tell-tale smell of leaking gas) it all came down. Miss Hodge's disappointment reflects our own disappointment that a single event like a kitchen gas leak should lead to such disproportionate consequences. She was fortunate to have survived the collapse dazed and slightly burnt – but four people died in their beds lower down the building as the concrete floors progressively collapsed.

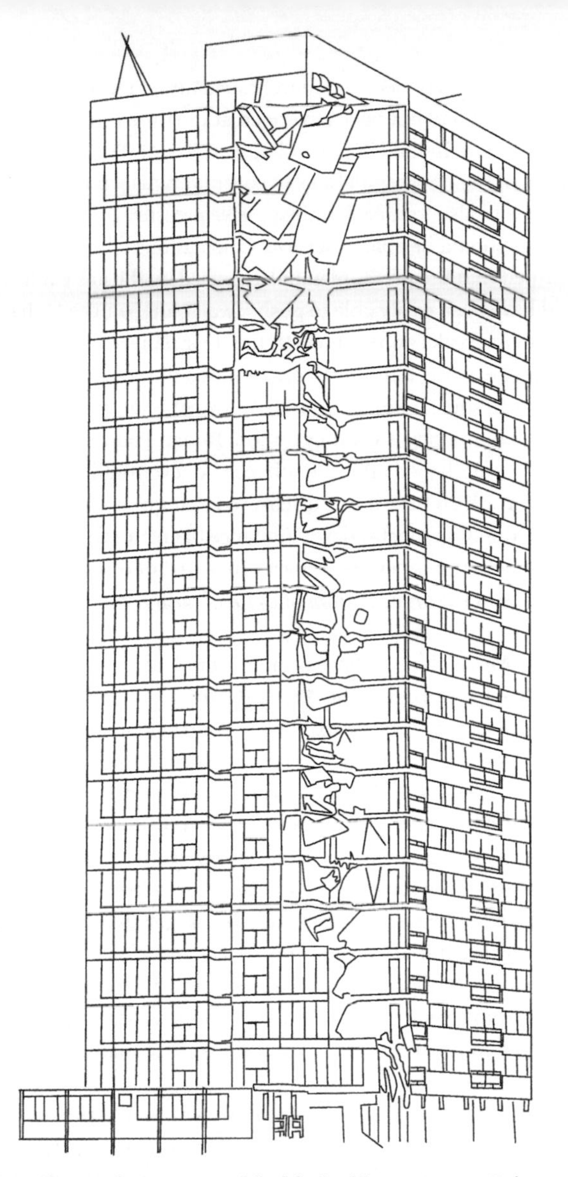

27. The collapse of one corner of the block of flats at Ronan Point was an example of progressive failure and the absence of structural robustness

Ronan Point can be seen as a purely structural failure, a mode of response that had not been envisaged at the time the building had been designed. In the UK, the collapse at Ronan Point led to changes in building regulations to cover both disproportionate collapse and also to specify the explosive pressure increase which buildings should be able to survive. Calculations after the collapse confirmed that the pressure impulse caused by the gas explosion would have been easily able to push a wall panel on the 18th storey out from its weak connection with the floor panels above and below. Owners of existing buildings were required to re-examine their designs – in some cases retrospective strengthening was feasible, in others demolition was the only option. Ronan Point itself was initially strengthened after the collapse but eventually demolished in 1986, at which time it was discovered that quite a lot of the actual construction was substandard. Vigilance to ensure that the quality of construction meets the specification is part of the essential *robustness* of the *organization* of the *realization* stage of design.

Timothy McVeigh must have been rather pleased with the extent of the damage caused to the Alfred P. Murrah Federal Building in Oklahoma City by a single explosion of a truck packed with explosive on 19 April 1995. The organizers of the aircraft attacks on the twin towers of the World Trade Center on 11 September 2001 would have been similarly impressed by the extent of damage to neighbouring buildings as well as the total collapse of the two 110 storey towers. In both cases the volume of building damaged or destroyed was considerably larger than the volume affected by the initial blast or impact. Could a closer attention to the possibilities of progressive collapse resulting from the transfer of loads from structural elements that had been directly eliminated have saved more of the buildings and their occupants?

On 28 July 1945, a B25 bomber aircraft was flown (presumably unintentionally – cloud level was very low) into the 79th storey of the Empire State Building in New York. Both engines became detached from the plane, one travelled right through the building and out the other side, the other went down a lift shaft. The steel structure of the building was so stiff that the effect of the aeroplane impact was rather small and damage was localized. The towers of the World Trade Center were designed with a much lighter structural form and the aircraft that crashed into them were considerably heavier than the B25. The fires resulting from the ignition of the aircraft fuel caused the steel floor trusses to soften and deform. The peripheral columns which were carrying the vertical loads from floors above were pulled in and then buckled and collapsed. The destruction of the fire protection to inner columns around the core of the buildings containing lifts and staircases also encouraged further structural damage from the heating. The towers survived the impacts but not the massive fires. But safety of a building under use, including adequate evacuation routes, is an expected element of its performance.

Heathrow Express collapse

In the 1960s, London Heathrow Airport (the busiest in the world at the time) required a tunnel to connect two areas separated by active runways in order to be able to move cargo on the airside of the airport. There was no possibility of closing the runways for the period of construction. The vertical alignment was constrained by the confined geometry of the site at each end which determined both the length of the tunnel and the complexity of the approach road network. The material through which the tunnel was to be constructed with an internal diameter of 11 m, was a rather homogeneous London clay. While there was about 7 m of ground cover between the tunnel and the runways there was only potentially 1.2 m cover within the clay – the overlying water-bearing sands and gravels had to be avoided. This ratio of cover to tunnel diameter was regarded as daringly low – and by some dangerously low. The tunnel was excavated using a circular shield which was really just an evolution of the shield devised by Marc Isambard Brunel for construction of the Thames tunnel in the 1830s. The clay revealed by the advance of the shield was supported by a flexible lining ring of concrete segments, allowing the cross-section of the tunnel to change modestly, treating the ground and the concrete segments as partners in the support strategy for the tunnel (Figure 28). Even in a modestly deformed shape the lining ring has a smooth continuity which allows it to carry a circumferential force around the tunnel. Tunnelling inevitably requires removal of ground from the face with a tendency for the ground above and ahead of the tunnel to fall into the gap. The success of the tunnelling operation can be expressed in terms of the volume loss: the proportion of the volume of the tunnel which is unintentionally excavated causing settlement at the ground surface – the smaller this figure the better. The tunnel as built produced settlements no greater than 11 mm and the volume lost was a mere 0.25 per cent, thus confounding the sceptics. The airport continued operating throughout the period of construction. This must be seen as a successful civil engineering project.

28. The concrete lining of the Taipei metro is erected ring by ring as the tunnelling machine bores its way through the ground

By contrast, on 21 October 1994 a large hole opened up in the middle of Heathrow Airport as a result of a collapse that had occurred in the tunnels that were under construction at a depth of some 30 m for the creation of a subterranean station for the mainline Heathrow Express rail link from London Paddington to the Airport. The station tunnels were being constructed using a technique sometimes known as the 'new Austrian tunnelling method' (NATM) in which the clay in the tunnel was excavated in sections with each section being provided with temporary support using steel mesh and sprayed concrete until the full final cross-section of the tunnel could be created and a permanent lining installed. For a trial section of the Heathrow Express tunnel constructed away from the crucial parts of the airport, the volume loss had been around 4 per cent. This trial was carried out precisely to review the potential for unacceptable movements in the ground when NATM was used. The elements of NATM had been used for rock tunnel construction for many years around the world before the Austrians chose to 'nationalize' the process. The essence of NATM is that support is provided for the newly exposed ground in response to the *observed* deformation of the tunnel. In rock tunnels, the deformations are usually small and develop slowly. It is quite feasible to plan a range of different quantities of support which can be mobilized rapidly. Experience in tunnelling in other ground conditions using this reactive technique has not been so positive.

Station tunnels are very three-dimensional with access tunnels parallel to the running tunnels, with cross passages between the various tunnels, and escalator tunnels leading to the ground surface. Such intersecting three-dimensional tunnel geometries are inevitably more risky than two-dimensional tunnels of constant cross-section because the symmetry of the excavation is being broken and the lining ring which is very strong when it is complete and intact is weakened (like the incomplete egg-shell).

How can failure of the tunnel be avoided? One route to reassurance will be to perform numerical analysis of the tunnel construction process with close simulation of all the stages of excavation and loading of the new structure. Computer analyses are popular because they appear simple to perform, even in three dimensions. However, such analyses can be no more reliable than the models of soil behaviour on which they are based and on the way in which the rugged detail of construction is translated into numerical instructions. Fully three-dimensional analyses are expensive and time consuming to perform and interpret.

Whatever one's confidence in the numerical analysis it will obviously not be a bad idea to *observe* the tunnel while it is being constructed. Obvious things to *observe* include tunnel convergence – the change in the cross-section of the tunnel in different directions – and movements at the ground surface and existing buildings over the tunnel. All these things were indeed being *observed* at Heathrow together with movements of markers within the ground over and beside the tunnel; but *observation* is not of itself sufficient unless there is some structured strategy for dealing with the *observations*. At Heathrow, unfortunately the data were not interpreted until after the failure had occurred. It was then clear that significant and undesirable movements had been occurring and could have been detected at least two months before the failure. There was a lack of robustness in the *organization* of construction and a lack of appreciation of the reasons for the *observations*.

In fact, the Heathrow collapse was not the only tunnel collapse to have occurred in the 1990s while NATM with sprayed concrete support was being used for tunnel construction in clay. Similarly serious collapses occurred in 1993 during the construction of metro tunnels in clay in Sao Paolo, Brazil, in 1994 in Munich, and in 1991 and 1994 in Korea.

The so-called new Austrian tunnelling method can be regarded as a member of a class of *observational* methods of design in which the uncertain prior knowledge of the ground is accommodated by having a range of design solutions available which can be applied without delay as the actual response of the ground is observed, and the 'ground model' can be updated – for better or worse. In the context of tunnelling this would mean having a strategy of different degrees of support for the tunnel depending on the *observed* tendency of the exposed soil or rock to break up or move. It is obviously important that the solutions for dealing with conditions worse than expected should be available for immediate application. It is also clear that such a design procedure should only be used for situations which are *robust* and not *brittle* – that is, not likely to fail suddenly if conditions are unexpectedly but not implausibly bad. There must be the possibility of a considered (not rushed) response to the observations.

There are analogies between the uncertainty in the ground and the behavioural uncertainty of organizations involved in civil engineering construction. For many of the well-known failures that have occurred the final technical straw that causes the failure (the strength of the material is exceeded at a sufficient number of points that a mechanism of failure is able to develop) is accompanied by some organizational dysfunction. Measurement (*observation*) of organizational dysfunction is rather more difficult than the measurement of the settlement of a building or convergence of a tunnel. Qualitative scales could be envisaged assessing the state of morale, communication between the parties involved in construction, technical confidence, technical competence, and so on, of the team – with appropriate ideas for improvement available. The need for robustness of the components of the process of *realization* of the recovery of the Heathrow Express project was taken extremely seriously to ensure that all parties involved were clear about the potential risks and about the importance of *observation* of all contributory elements.

A successful civil engineering project is likely to have evident *robustness* in *concept*, *technology* and *realization*. A *concept* which is unclear, a *technology* in its infancy, and components of *realization* which lack coherence will all contribute to potential disaster.

Minneapolis Interstate 35W highway bridge

Interstate 35W passes through the middle of Minneapolis crossing the Mississippi River on a steel trussed bridge, constructed in 1964–7. Although the bridge was rated as 'structurally deficient' in inspections in 2001 and 2005 its strengthening was not regarded as a high priority. However, on the morning of Wednesday 1 August 2007 the central span of the bridge collapsed suddenly: 13 people were killed. Investigation after the collapse revealed that the failure was probably triggered by fatigue failure of a gusset plate linking together several of the girders.

Fatigue is the term used to describe a failure which develops as a result of repeated loading – possibly over many thousands or millions of cycles. The crashes of Comet jet airliners in 1954 were discovered to be the result of crack growth from the corner of a window under the repeated cycles of loading caused by take-off, cabin pressurization, in-flight loading, and landing. Fatigue is the central theme of the Neville Shute book *No highway*, published in 1948 at a time when the phenomenon of fatigue – especially important for aircraft structures – was beginning to be understood.

Fatigue cannot be avoided, and the rate of development of damage may not be easy to predict. It often requires careful techniques of inspection to identify the presence of incipient cracks which may eventually prove structurally devastating. Ultrasonic techniques may be able to detect small defects; the presence of a crack will tend to lower the stiffness of the structural element in which it is located and hence lower its natural frequency (just like tuning a piano). A highway bridge is subjected to various sources of repeated loading: daily and seasonal cycles of temperature variation as well as the dynamic loading from the axles of eight lanes of traffic. But the essence of robust structural design is that the loss of one element of a structure should not produce disproportionate consequences. In this case, the failure of one gusset plate led to a hurried attempt by the bridge structure to find other ways of carrying the loads, but the structure rapidly unzipped until its failure was inevitable. The failure was the result of the loss of strength of one element of the structure but it was also the result of a lack of appreciation of the non-robustness of the design and thus of the enhanced importance of regular careful inspections. Naturally, after the Minneapolis collapse many other bridges of similar design were subjected to particularly careful inspection.

Tohoku earthquake 2011

Robustness of design becomes quite a broad concept which has to be concerned with the complete system – structural integrity, human behaviour in an emergency, building services (ventilation, smoke removal, information systems). Redundancy (duplication) is needed in all areas in order to withstand and contain moderate amounts of damage. This applies equally to a tall building under impact from aircraft and to a nuclear power station under a supposedly controlled emergency simulation.

The Fukushima nuclear power station in the Tohoku region of north-east Japan was badly damaged by the magnitude 9 earthquake and subsequent tsunami that devastated the area on 11 March 2011. Initial reports of the consequences of the earthquake concentrated on the general destruction caused by the combination of these two major natural disasters and there are many pictures of the devastation. However, as information about events at Fukushima emerged, it triggered reviews of nuclear safety around the world and released a wave of, not obviously rational, opposition to any future development of nuclear power.

That there was significant leakage of radioactivity might be put down to 'bad luck'. After all, the probability of both a major earthquake *and* a major tsunami occurring at the same time might appear low. The earthquake automatically triggered a shut down of the three reactors of the six that were actually operating. Two of the six were closed for routine maintenance but all had fuel rods in place and all required continued cooling. The power station was protected by a 6 m high sea wall but this was insufficient to provide protection against the 14 m tsunami which arrived less than an hour after the earthquake. This flooded the generators and pumps and destroyed the emergency electrical supply and the connection to the external electrical grid. Without the pumps, the water cooling system could not be kept going and the radioactive cores overheated. Partial melt-down occurred in three reactors, fuel rods that should have been submerged became exposed because the water was boiling away, and holes formed in the base of the pressure vessels as a result of the high temperatures – as high as 2800°C within a few hours of the earthquake. Explosions occurred in the containment structures because of hydrogen leakage from the pressure vessels. There were delays in setting up an emergency seawater cooling system – absence of power, difficulty of access because of earthquake damage to roads — and it was known that the introduction of salt water would severely damage the reactor.

But 'bad luck' is not acceptable for sensitive installations such as nuclear power stations. And, of course, the two events are not uncorrelated (tsunamis are usually caused by earthquakes) so the probability that the two events could occur together must be considered quite high. The reactors at Fukushima are boiling water reactors designed by General Electric, which were brought into operation between 1971 and 1976. Given that the continuity of water cooling is essential for safe operation of the reactors, this should evidently be a central element of any overall plant safety plan. The possibility that all supplies of power for the coolant pumps could be destroyed is so serious that the design thresholds (earthquake acceleration, tsunami wave height, . . .) need to be set high.

But there are also suggestions that the regulatory environment at Fukushima was not entirely satisfactory. In 2002, the operating company, Tokyo Electric Power Company, admitted to having falsified safety records at Fukushima. There are suggestions of 'regulatory capture': where the regulator, supposedly acting in the public interest, advances the particular interests of the commercial sector that it should be regulating; and senior regulators take up highly paid posts in the industry that they were previously regulating.

It is certain, anyway, that the systems at Fukushima were insufficiently robust to cope with the disaster.

Contract

> *Fasolt*: Soft sleep closed your eyes, while we were working to build your hall. Working hard, day and night, heavy stones we heaped up high; lofty towers, gates and doors, guard and keep, your castle walls secure. There stands what you ordered, shining bright in morning light. There's your home; we want our wage!
> *Wotan*: You've earned your reward; what wages are you asking?
> *Fasolt*: The price was fixed, our bargain was made; have you so soon forgot? Freia, the fair one, Holda, the free one – your hall is built and Freia is ours.
> *Wotan*: Plainly your work has blinded your wits. Ask some other wage: Freia cannot be sold.
> *Fasolt*: What's this now? Ha! Breaking your bond? Betraying your word? On your spear shaft, read what is graved; would you dare to break your bargain?

In Wagner's opera, *Rhinegold*, the head of the gods, Wotan, has contracted with the giants Fasolt and Fafner to build his fine new palace, Valhalla. Naturally, having completed the construction as specified, they expect him to keep his side of the contract. Wotan tries to wriggle out of the contract, and the remainder of the *Ring* cycle of operas describes the consequences – in the end Fasolt and

Fafner (among others) have been killed and Wotan (with the other gods) is destroyed along with his palace, Valhalla.

The need for properly drawn up contracts between the various parties involved in civil engineering projects is obvious: there are standard forms available. Lawyers have no trouble in spotting clauses which have been breached. Fasolt and Fafner have an open and shut case against Wotan which they then proceed to weaken by making a verbal agreement to an alternative payment. A god's word is not necessarily his bond. For most of the projects that have been described we have seen a combination of a *concept* (an architectural scheme or a client outline) and an enabling *technology* (which might be detailed material understanding, or prior experience, or a daring visionary leap) and a *realization* (bringing together the several parties involved in design, financing, and construction). The division of responsibilities seems to lend itself to strict separation of contractual obligations. One suspects that Wotan gave the giants a pretty free hand to build his palace – the general problem definition was his, the *technology* and *realization* were left to them. The giants accepted a fixed price contract and it was up to them to look for ways in which they could save time and resources in order to maximize their profit while still meeting the specification that Wotan had given them. Perhaps, as in many complex publicly procured projects, there would have been constant changes to the detail of the design requirements which would have led to potential claims by Fasolt and Fafner for reimbursement beyond the original fixed price. But the possibilities of such claims are overtaken by operatic events.

The balance of responsibilities and assignment of risk has moved around through the ages. The three principal parties are the *client* who wants the project (and possibly a separate financier providing the necessary funding); the *designer* (engineer) who has to interpret the needs of the client into a form that will satisfy those needs and that can actually be constructed; and the *constructor* (or *contractor*) who will turn that design into the physical reality. Some of these roles may be combined: the Brunels were essentially both *designer* and *constructor* for the Thames tunnel since they not only knew exactly how the tunnel should look and be constructed but also engaged the labourers for that construction. Rennie was fully aware that, having designed the Bell Rock lighthouse, it was he who carried the risk of his chosen design failing to meet the requirements of the Northern Lighthouse Board; Stevenson was both design representative (*resident engineer*) and also manager of the construction itself (*site agent*). If failure had occurred as a result of poor workmanship rather than a fault in the design he would have had to assume some responsibility for not having taken adequate steps to assure the quality of the work. When things do go wrong – when costs escalate or construction methods have to be changed because the client has changed his mind or because the ground conditions are not quite what had been assumed – then insurance companies and lawyers try hard to ensure that the blame and the financial consequences do not end with their clients. Certain forms of contract may be more helpful than others. And in the present litigious world it is usually only the lawyers who can be guaranteed to win.

Civil engineering designers can enter the project organization at two distinct stages. The *client* will need to have some idea about the cost of the project before seeking a price from the *constructor*: a design which can turn the concept into reality is necessary. But if the full responsibility for realizing the project is passed to the constructor then it is reasonable for him to want to introduce his own designer and to seek more economical ways of meeting the client's requirements: another design (and another designer) is required. A design and build contract merges the two design stages under the responsibility of the constructor. The client believes that he is proceeding with the project under full certainty of cost and schedule, and that all risk is placed firmly with the constructor who will price the work accordingly.

A turnkey project takes this further: the client engages a project management company which handles the supervision of all aspects of the design and construction and then hands to the client a fully operational facility. Such an organizational structure is attractive for complex projects such as the construction of a nuclear power station, or a major river crossing, or a chemical refinery, or a substantial stretch of railway line or motorway, where specialist subcontractors will come from a wide range of disparate areas. The cost will be set accordingly.

At the other extreme would be unit price contracts based on a list of the quantities of materials, labour, and equipment that will be required for the project. The basis for the estimate of the project price is clear but there is no incentive to seek more economic alternatives, and the constructor, by subtly raising the mobilization costs incurred in having equipment on site and reducing the actual unit costs of performing the work with this equipment (or vice versa, depending on his interpretation of the project) will naturally seek to improve his profitability.

As in many other areas of public life there is a conflict between the degree of trust and confidence that each party has in the other parties engaged in the project, and the detail of the liabilities actually specified in the contract – 'passing the buck' seems to be a natural human reaction to adverse events. The formal sharing of the consequences of uncertainty – positive or negative – will be largely a matter of contractual obligations. An organizational structure which assumed that everyone shared the common goal of an economically and functionally successful completion and which shared the risks and the benefits of cost savings among the partners would be attractive.

Can strict separation of the roles and assignment of the risks be justified? A client who wants financial certainty will try to place all possible risks onto the constructor. But having assumed all these risks, a constructor will price his work accordingly – if the risks do not materialize then the client will have ended up paying much more than necessary and the profit of the constructor will be correspondingly larger. A complete formal separation of designer and constructor removes the continuity of design which must be desirable in ensuring that the person building the project is fully aware of the assumptions that underpin the design and is continually looking for evidence of the possibilities that some of these assumptions may prove to have been violated.

Division of contractual responsibilities obviously implies corresponding liabilities. In that there remains a significant element of empiricism and experience in much civil engineering, there must be an associated route to making it a learning profession, learning from past failures to avoid a repetition of mistakes. Too often the concerns of the insurance companies and the lawyers working with the different parties result in settlements or apportionment of damages following a failure being made away from the public eye – thus losing, possibly for ever, the learning potential which exists quite independently of the legal decisions.

Some projects would clearly be regarded as failures – a dam bursts, a flood protection dyke is overtopped, a building or bridge collapses. In each case there is the possibility of a technical description of the processes leading to the failure – in the end the strength of the material in some location has been exceeded by the demands of the applied loads and the load carrying paths have been disrupted. But failure can also be financial or economic. Such failures are less evident: a project that costs considerably more than the original estimate has in some way failed to meet its expectations. A project that, once built, is quite unable to generate the revenue that was expected in order to justify the original capital outlay has also failed. Is the Sydney Opera House a success or a failure? It took 15 years to be built and cost 10 times the original estimates and has a shape and layout that in many ways hinder its function as a multiple centre for performing arts. On the other hand, it created an internationally instantly recognized icon which has had a huge public relations and touristic benefit for the city of Sydney. That uncostable benefit was not one of the criteria of the original design specification. Is the Channel Tunnel a success or a failure? It too cost much more than originally expected and took longer to build. For a whole range of reasons it has not been able to generate the income necessary to cover the costs of the original finance. There have been two serious fires in the tunnel which have severely weakened the tunnel lining and possibly come close to producing serious inundation. Is the Hoover Dam (or the Mangla Dam on the Indus River in Pakistan, or the Aswan Dam on the Nile, or the Three Gorges Dam in China) a success or a failure? The Hoover Dam has not collapsed, so technically it is a success. It has provided water for irrigation of land which would otherwise be desert – but in so doing has removed water from riparian farmers downstream where the agriculture was perhaps more sustainable in the long term. And the long distance pumping has demanded enormous amounts of energy. The silt which would originally have been carried down the Colorado River is held up behind the dam – and other dams on the river – so that the capacity of the retained reservoir has reduced. Long term irrigation leads to problems with salination of the soil, and the greatly reduced flow in the downstream river has a salt content which is too high for many of the uses to which the water would previously have been put. And one man's reservoir (power generation, irrigation, flood prevention) is another man's environmental despoliation. Engineering is always about choices – usually political choices – which imply the comparison in a single equation of quite incompatible and unmeasurable quantities.

Chapter 6
Civil engineering: looking forward

The future in the past

What would the early 18th century engineer have forecast as the challenges facing engineers (civil engineers) over the succeeding century? The Industrial Revolution was moving forward, people were migrating from the clean air of the country to jobs in smoke-filled towns, scientific knowledge was increasing rapidly, new materials were emerging leading to new structural forms. He (all the engineers that we know about were men) might have predicted that problems of water supply and waste disposal would arise in rapidly expanding cities; that better means of communication and transport of people and goods would be needed; that with a reduced agricultural labour force international trade would be required for food and other materials not readily available locally; and that docks and harbours would be required for the raw inputs and finished outputs of the factories that had sprung up on the back of the Industrial Revolution. Perhaps he might have spotted that static steam engines could be adapted for self-propulsion. Civil engineering developed solutions in all these areas - perhaps more rapidly than had been anticipated a century before.

What would our forecasts contain now in the early 21st century?

Glimpses

Traffic in towns

In 1963, a report entitled *Traffic in towns* was produced for the UK Minister of Transport by a working group led by Colin Buchanan. It was produced at a time when motor transport was growing very rapidly (in the UK) and travel by car was taking over from public transport by rail or road. Many of the messages are valid today – the illustrations of the conflicts between pedestrians and vehicles and between cyclists and vehicles, and of the urban clutter associated with parking of cars look very familiar, only the appearance of the cars and the styles of personal dress look rather dated.

The concept with which Buchanan emerges is hardly surprising (and he admits that it is hardly original). It is one of creating areas of pleasant environment where people can live, shop, work, and play without being affected by urban traffic. Equally there must be a network of roads which distribute traffic though the town and to the person-centred environments. Having thought through the problem in the abstract, Buchanan describes a number of case studies in which existing UK towns are analysed in order to understand the nature of the traffic flows. Suggestions are made as to how appropriate traffic management and provision for separation of cars and pedestrians could dramatically improve the environment. The last of these case studies deals with a block of central London and tests three solutions – one for complete redevelopment, the others for partial and minimum redevelopment. Complete redevelopment puts all the traffic at ground level with the pedestrians weaving their way among the somewhat anonymous tower blocks on a raised podium; the solution to a traffic problem involves more than just tampering with the roads. Minimum redevelopment is more subtle because it has to work with the existing road layout and recognize the finite available capacity for traffic handling.

The general ideas proposed by Buchanan were being applied in the new towns created in Britain after the Second World War (and in the 'garden cities' proposed by Ebenezer Howard towards the end of the 19th century influenced by Frederick Olmsted's urban parks in the USA). But the planners failed to predict the demand for car ownership – a car was no longer a status symbol but something that (almost) everyone had. Initial plans for one garage for every four households moved towards an estimate of 1.5 cars per household. This was a density of ownership which Buchanan reckoned virtually impossible to incorporate into the remodelling of an existing town. He ends with a description of Venice which, par excellence, separates the distribution network of canals from the pedestrian network. It is this total separation that he extracts as the message for a well-functioning city and which was applied to the total redevelopment of the central London block.

Management of traffic is just one of the aspects of infrastructure which needs to be controlled in order to preserve a sustainable quality of life. Many aspects of the solutions to the problems of traffic and transport are pure civil engineering; the interface with town planners and traffic consultants is another one across which civil engineers must be able to communicate.

Dongtan – eco-city

Traffic in towns was looking for ways in which city life could cope with the needs for mobility of the residents. Sustainability may not have been the word of the moment but there was certainly a concern for the environment and comfort of daily life. Buchanan aimed to live with the automobile. The ambitious plans for the eco-city Dongtan aimed to exclude it completely.

Dongtan was to be built on an island in the mouth of the Yangste river alongside the city of Shanghai. It was conceived in 2005 as a prototype of a new style of development for China; a city with zero-greenhouse emission public transport, no cars, complete self-sufficiency in water and energy, and energy efficient buildings. Designs were prepared by Arup on behalf of the Shanghai Industrial Investment Company with the intention of housing a population of 10,000 in time for Expo 2010 and 500,000 by 2050. Much innovation has gone into the designs of detailed elements of the city in order to meet challenging ecological targets, particularly those concerning energy generation from renewable resources; and waste elimination. All housing would be within 7 minutes walking distance of public transport; food would be produced on the island; energy consumption of buildings in Dongtan would be about one third of that for conventional buildings. Sustainability for a city requires social, economic, and cultural sustainability as well as environmental sustainability.

However, it seems that in the end there has not really been the serious *political* and *financial* will to push the project ahead. The *concept* is there, the *technology* is being developed – we can be sure that some of the ideas will see application elsewhere – but the elements of *realization* are absent. The *concept* of an eco-city is widely attractive in the abstract but less attractive in reality to those who are being expected to purchase dwellings and live there. The bridge-tunnel from Shanghai to Dongtan has been built, opening the way to development of a style familiar in Shanghai and perhaps more popular with purchasers than the low energy alternatives.

Infrastructure and investment

On 3 December 2011 *The Economist* included an article headed *Weapons of mass construction*:

> Giant steel columns are being drilled 16 metres into the earth at Royal Oak in west London, preparing the way for twin 6.2 metre wide tunnels to be carried under ground. By 2018 this muddy patch will host Crossrail, a new east–west commuter service which will bring another 1.5 million people within 45 minutes of central London. At a cost of £14.5 billion ($22.5 billion) the biggest engineering project in Europe is just one of 40 programmes labelled "nationally significant" [in] the British government's National Infrastructure Plan [which] also lists 500 further

> proposals from the private and public sector, estimated at £250 billion worth of work in total…

There is a bit of journalistic dramatization in the description of the engineering but this introductory paragraph contains several relevant details. Development of infrastructure does not stop (Figure 29) – we expect governments to ensure that our infrastructure remains up to date and effective. Provision of infrastructure (whether new or maintenance of old) requires significant investment; at a time of economic recession investment in infrastructure is seen as a way of giving the economy a short term boost. Improvement of the rail transport infrastructure should improve the quality of life in the London urban area and reduce, to a small extent, the need for energy consumption by the use of personal motor vehicles. The Crossrail project is underground – out of sight, like much of the infrastructure that we take for granted. In a crowded urban environment we have to take our engineering downwards. But the ground beneath London is already well exploited with existing underground railways, sewers, water and gas pipes and other communication links – not to mention the foundations for the buildings, old or new, that are present at the ground surface. Any new tunnels (and stations and escalator tunnels) must not only avoid this existing infrastructure but also ensure that it is not damaged through ground movement or ground vibration associated with construction. It should be no surprise that there are significant technical challenges associated with the 'biggest engineering project in Europe'.

We can set this description of 'work in progress' alongside the two visions of the future, one (Buchanan) working in small steps with the realities of the current situation, the other (Dongtan) starting idealistically from a clean canvas (once the existing agricultural use of the land has been cleared). The future remains difficult to predict; human nature is as unpredictable as any natural phenomenon.

29. Heathrow Airport Terminal 5: building for the 21st century?

Advances

Analytical tools

Man's ability and ingenuity to develop new and smaller and more effective electronic devices appear to know no bounds. Miniaturization of integrated circuits reduces the distances between components and increases the speed with which they can exchange information. As one consequence, computing speed and availability of computing resources have increased dramatically over the past few decades and will no doubt continue to increase. The power that was available from processors that occupied a complete room in the 1960s is now contained within devices that can be held in the hand. We can analyse, simulate, and control systems of a complexity that we could not previously contemplate. There is of course an attendant danger that the users of any information-processing device for analysis or control will place greater trust than can be justified in the electronic output. For the purposes of civil engineering design there is an essential parallel step of trying to obtain a confirmatory result using a back-of-the-envelope calculation. However, for control of a complex system such as a nuclear power station it is harder to be confident that there is no unthought-of circumstance that could occur to upset our assumptions and disrupt our best laid plans for safe operation.

Sensors and communication

Communications have become faster and more ubiquitous; information can be transmitted almost anywhere around the world in an instant. Data required for analysis or design of a project or for control of a completed civil engineering system can be processed remotely and control achieved from a distance. Developments in the miniaturization of electronic devices have produced sensors which can be embedded in structural elements to report local displacements, stresses, accelerations, and temperature as part of an active programme of condition monitoring – which only makes sense if it is part of a coherent programme of asset management. There has to be a recognized purpose for the monitoring. For example, small amounts of corrosion or structural damage or foundation degradation will lead to changes in the dynamic response of structural systems. There is generally sufficient low level environmental background energy to enable this changing response to be detected automatically (and remotely) so that the source can be identified (perhaps by numerical simulation using the evolving computer power) before a bridge or building or dam or offshore structure actually becomes unsafe. Robustness of the structure and the monitoring system is obviously essential. Automatic response to feedback has the attraction of elimination of the fallible human being but leaves us in a dehumanized world with increasing reliance on machines designed and programmed by those same fallible humans. The machines have to be programmed to look for the unknown and unexpected – something which the human brain can often do rather well.

Innovative materials

If there is the possibility of monitoring the occurrence of damage then there may also be the possibility of automatically taking action to *self-heal* the damage by chemical or biological means: a goal of intelligent systems. The ground, for example, is naturally full of microorganisms some of which, given the right chemical environment, can be used to precipitate carbonate gel which can bond particles of sand and provide a little surface strength, sufficient to delay loss of strength through liquefaction in an earthquake. All the microorganisms need is the appropriate nutrition – and tender loving care. There is definite potential for making beneficial use of 'intelligent bugs'. Some bugs and fungi are particularly suited to cleaning up toxic chemicals and helping to turn brownfield into greenfield sites and thus helping to slow the encroachment of development into green and pleasant countryside. Engineering requires lateral thinking – the best solutions may well be surprisingly non-traditional ones which require civil engineers to learn the language of new groups of collaborators. Life itself, at all scales, can be seen as one of the 'great sources of power in nature'.

New destinations

The destination for our civil engineering projects may be somewhere far away in space or on another planet. Gravity will still be a stabilizing force but perhaps not as significant as on Earth. Civil engineering in space will challenge our inventiveness for lightweight, compact, collapsible, deployable structures. Transport costs (and times) will be high until we have discovered ways of exploiting local planetary mineral resources. We should be cautious about dismissing the possibility of creating and supporting human settlements in space, and relish the excitement of searching for novel engineering solutions.

There are also new and challenging destinations for civil engineers on this planet. Going underground to find space for transport and other infrastructure avoids despoliation of the visible urban or rural environment but has to cope with the uncertain geological conditions and with the foundations and tunnels left by others who have been there earlier. As prices of minerals – metals, semiconducting materials, hydrocarbons – increase so the economics of exploration and production change. Engineering for extraction moves into more hostile areas – Arctic, Antarctic, deep oceans – with corresponding challenges and responsibilities!

Sustainability

Sustainability is one of those words with which we are regularly accosted in the media. We recognize that it is a good thing while perhaps not being quite sure what the implications might be for us. Sustainability might be defined as satisfying our current economic, social, and environmental demands without jeopardizing the possibility that future generations may be able to enjoy a quality of life no worse than ours. That definition contains some rather vague and unmeasurable terms but implies a degree of personal social responsibility and sacrifice for the general future good of society. It certainly implies making choices between apparently irreconcilable alternatives.

We are encouraged to think about our *carbon footprint* – the amounts of carbon dioxide and methane that are implicitly generated by our lifestyle. While we are directly aware of the fuels we use for transport and heating, we may be less aware of the energy and carbon consumption that has contributed to the production of the materials and foodstuffs and products that we like to have around us. These greenhouse gases are important because they contribute to the *greenhouse effect* forming a layer in the atmosphere which reflects heat radiated from the Earth, leading to *global warming* and *climate change*. This is a controversial area – the anthropogenic rise in temperature (caused by human activity) is superimposed on changes which occur naturally over the millennia. The prediction of the effects of increased quantities of these gases in the atmosphere on climate in the decades and centuries ahead is very sensitive to the details of the numerical models that are used to simulate the entire Earth system – and we know how difficult it is to predict with accuracy the weather tomorrow.

There are things that contribute to our present quality of life which we would be reluctant to lose. We happily jump into the motorcar to go to the supermarket to buy vegetables that have been shipped halfway round the world. We expect to be able to use dishwashers, washing machines, tumble driers, heaters, showers whenever we want to and would be disappointed if energy (electricity or oil or gas) or water were only sporadically available – but the increasing introduction of energy generation schemes based on renewable sources such as solar or wind or wave which are not always able to function when we need them is leading to new ways of energy distribution and management and storage: the 'smart grid'. As the populations of the developing world develop they too will expect to have the same access to energy and water and consumer products – global demand will increase. Having for many years watched enviously on television and film the quality of life and possessions in the developed world they may well be reluctant to skip this stage of civilization and espouse a more austere lifestyle simply for the probable future benefit of the planet.

Civil engineers are the servants of society. They will be expected to respond to the challenges that arise and to provide protection against increasingly severe weather and changing sea level. They will be expected to play their role in the provision of safe and reliable means of generation of electricity whether from water power, or coal, or oil, or nuclear reactors, or offshore wind, or tidal or wave energy, or the sun. Decisions and choices are made by politicians who are concerned with popularity and re-election, which may well be in conflict with rational debate. Too often engineers remain below stairs, seen but not heard. Insistent engagement in informed discussion and effective communication of the real technical choices are parts of the role of the responsible civil engineer. We can do much more.

We are unsure of the pace of the changes and of their likely effects. Man may be unwilling to accept the proposed harnessing or husbanding of the sources of power for his use and convenience. Exactly how (and where) we will provide the infrastructure, and satisfy expectations for safety when confronted by a changing natural environment is difficult to forecast: it is the role of engineers to rise to such challenges. Success will come to those who are prepared to think laterally, to speak out, and to study disciplines which are not traditionally seen as part of civil engineering, looking for innovative ideas which are capable of engineering application.

Sources

Introduction: civil engineering

The general advancement of mechanical science...
Royal Charter of the Institution of Civil Engineers (1828)

broadbrush way of describing...; Architecture lacks the linearity...; Once we enter into...
Architect and engineer: a study in sibling rivalry, Andrew Saint

Chapter 1: Materials of civil engineering

Before the Roman came to Rye...
G.K. Chesterton

Chapter 2: Water and waste

Earth has not anything...
Upon Westminster Bridge, (3 Sept 1802) William Wordsworth

Chapter 3: 'Directing the great sources of power in nature'

With the civil engineer...
Records of a family of engineers, Robert Louis Stevenson (1896) (unfinished)

Once did she hold...
On the Extinction of the Venetian Republic, (1802) William Wordsworth

It wasn't the miles...
The tin roof blowdown. James Lee Burke 2007 (Phoenix)

Chapter 4: Concept – technology – realization

The paramount importance...
Ove Arup: Masterbuilder of the twentieth century Peter Jones

Say not of me...
Poem XXXVIII, Underwoods Robert Louis Stevenson (1887)

stout Cortez...
On first looking into Chapman's Homer John Keats

Is then no nook...
On the projected Kendal and Windermere railway (12 Oct 1844)
William Wordsworth

Chapter 5: Robustness

Soft sleep...
Rhinegold, Richard Wagner, translated Andrew Porter

Chapter 6: Civil engineering: looking forward

Giant steel columns...
Weapons of mass construction (Economist 3 December 2011)

Further reading

If you want to pursue some of the themes of this book you may find these books of interest. Searches on the world wide web using as keywords the titles of projects to which reference has been made will lead to massive amounts of information – not all of it reliable, of course.

The gradual separation of science, architecture, and civil engineering over the past half millennium

Ferguson, Hugh and Chrimes, Mike (2011) *The civil engineers: The story of the Institution of Civil Engineers and the people who made it.* ICE Publishing

Holmes, Richard (2008) *The age of wonder: how the romantic generation discovered the beauty and terror of science.* HarperPress

Saint, Andrew (2007) *Architect and engineer: a study in sibling rivalry.* Yale University Press

Discussion of the ways in which civil engineering structures work (or do not work)

Blockley, David (2010) *Bridges.* Oxford University Press

Levy, Matthys and Salvadori, Mario (1992) *Why buildings fall down: how structures fail.* W. W. Norton

Salvadori, Mario (1990) *Why buildings stand up: the strength of architecture.* W. W. Norton

Detective work concerning the construction of the Pyramids and the Gothic cathedrals

Ball, Philip (2008) *Universe of stone: Chartres Cathedral and the triumph of the medieval mind.* Bodley Head

Fitchen, John (1961) *The construction of Gothic cathedrals: a study of mediaeval vault erection.* Oxford, Clarendon Press

Parry, Dick (2004) *Engineering the Pyramids.* Sutton Publishing

Snapshots of infrastructure in Pompeii, improvement of infrastructure in London, reclamation and its consequences

Beard, Mary (2008) *Pompeii: The life of a Roman town.* Profile Books

Darby, HC (1956) *The draining of the fens.* Cambridge University Press

Fletcher, CA and Spencer, DT (eds.) (2005) *Flooding and environmental challenges for Venice and its Lagoon.* Cambridge University Press

Halliday, Stephen (2009) *The great stink of London: Sir Joseph Bazalgette and the cleansing of the Victorian metropolis.* The History Press

Sustainability and environmental issues tend to stir controversy

Carter, Vernon Gill and Dale, Tom (1974) *Topsoil and civilization.* University of Oklahoma Press

Lomborg, Bjørn (2001) *The sceptical environmentalist: Measuring the real state of the world.* Cambridge University Press

Miller, Mervyn (2010) *English garden cities: An introduction.* English Heritage

Reisner, Marc (1993) *Cadillac desert: The American West and its disappearing water.* Penguin Books

(1963) *Traffic in towns: A study of the long term problems of traffic in urban areas.* Report of the Steering Group. London HMSO

A detailed account of a tunnel failure, how it might have been avoided, and how successful completion of the project was ensured

Health and Safety Executive (2000) *The collapse of NATM tunnels at Heathrow Airport.* HSE Books

Lownds, Sue (1998) *Fast track to change on the Heathrow Express.* Institute of Personnel and Development
Muir Wood, Alan (2000) *Tunnelling: management by design.* E and FN Spon

Construction challenges of the 15th and 20th centuries

Jones, Peter (2006) *Ove Arup: Master builder of the twentieth century.* Yale University Press
King, Ross (2000) *Brunelleschi's dome: The story of the Great Cathedral in Florence.* Pimlico
Murray, Peter (2004) *The saga of Sydney Opera House.* Spon Press

Adjusting the historical record?

Bagust, Harold (2006) *The greater genius? A biography of Marc Isambard Brunel.* Ian Allan
Mathewson, Andrew and Laval, Derek (1992) *Brunel's tunnel…and where it led.* Brunel Exhibition Rotherhithe
Paxton, Roland (2011) *Dynasty of engineers: The Stevensons and the Bell Rock.* Northern Lighthouse Heritage Trust
Vaughan, Adrian (1991) *Isambard Kingdom Brunel: engineering knight errant.* John Murray

Index

A

B

J

K

L

M

N

O

T

U

V

W

INNOVATION

A Very Short Introduction

Mark Dodgson & David Gann

This *Very Short Introduction* looks at what innovation is and why it affects us so profoundly. It examines how it occurs, who stimulates it, how it is pursued, and what its outcomes are, both positive and negative. Innovation is hugely challenging and failure is common, yet it is essential to our social and economic progress. Mark Dodgson and David Gann consider the extent to which our understanding of innovation developed over the past century and how it might be used to interpret the global economy we all face in the future.

'Innovation has always been fundamental to leadership, be it in the public or private arena. This insightful book teaches lessons from the successes of the past, and spotlights the challenges and the opportunities for innovation as we move from the industrial age to the knowledge economy.'

Sanford, Senior Vice President, IBM

www.oup.com/vsi

RISK

A Very Short Introduction

Baruch Fischhoff & John Kadvany

Risks are everywhere. They come from many sources, including crime, diseases, accidents, terror, climate change, finance, and intimacy. They arise from our own acts and they are imposed on us. In this *Very Short Introduction* Fischhoff and Kadvany draw on both the sciences and humanities to show what all risks have in common. Do we care about losing money, health, reputation, or peace of mind? How much do we care about things happening now or in the future? To ourselves or to others? All risks require thinking hard about what matters to us before we make decisions about them based on past experience, scientific knowledge, and future uncertainties.

www.oup.com/vsi

STATISTICS

A Very Short Introduction

David J. Hand

Modern statistics is very different from the dry and dusty discipline of the popular imagination. In its place is an exciting subject which uses deep theory and powerful software tools to shed light and enable understanding. And it sheds this light on all aspects of our lives, enabling astronomers to explore the origins of the universe, archaeologists to investigate ancient civilisations, governments to understand how to benefit and improve society, and businesses to learn how best to provide goods and services. Aimed at readers with no prior mathematical knowledge, this *Very Short Introduction* explores and explains how statistics work, and how we can decipher them.

www.oup.com/vsi